State of the World's Oceans

Michelle Allsopp • Richard Page
Paul Johnston • David Santillo

State of the World's Oceans

 Springer

Michelle Allsopp
Greenpeace Research Laboratories
School of Biosciences
University of Exeter
Exeter
UK

Paul Johnston
Greenpeace Research Laboratories
School of Biosciences
University of Exeter
Exeter
UK

Richard Page
Greenpeace
Canonbury Villas
London, UK

David Santillo
Greenpeace Research Laboratories
School of Biosciences
University of Exeter
Exeter
UK

ISBN: 978-1-4020-9115-5   e-ISBN: 978-1-4020-9116-2

Library of Congress Control Number: 2008936032

Cover Figure: © Greenpeace/Daniel Ocampo

Printed on acid-free paper

springer.com

Preface

The world's oceans cover 70% of the earth's surface and are home to a myriad of amazing and beautiful creatures. However, the biodiversity of the oceans is increasingly coming under serious threat from many human activities including overfishing, use of destructive fishing methods, pollution and commercial aquaculture. In addition, climate change is already having an impact on some marine ecosystems. This book discusses some of the major threats facing marine ecosystems by considering a range of topics, under chapters discussing biodiversity (Chapter 1), fisheries (Chapter 2), aquaculture (Chapter 3), pollution (Chapter 4) and the impacts of increasing greenhouse gas emissions (Chapter 5). It goes on to explore solutions to the problems by discussing equitable and sustainable management of the oceans (Chapter 6) and protecting marine ecosystems using marine reserves (Chapter 7).

Presently, 76% of the oceans are fully or over-exploited with respect to fishing, and many species have been severely depleted. It is abundantly clear that, in general, current fisheries management regimes are to blame for much of the widespread degradation of the oceans. Many policy-makers and scientists now agree that we must adopt a radical new approach to managing the seas – one that is precautionary in nature and has protection of the whole marine ecosystem as its primary objective. This 'ecosystem-based approach' is vital if we are to ensure the health of our oceans for future generations.

The ecosystem approach is one that promotes both conservation and sustainable use of marine resources in an equitable way. This is an holistic approach which considers both environmental protection and marine management together, rather than as two separate and potentially mutually exclusive goals. In this regard, the establishment of networks of fully protected marine reserves is paramount to the application of the ecosystem-based approach. Marine reserves act like national parks of the sea. They provide protection of whole ecosystems and enable biodiversity to both recover and flourish. They also benefit fisheries because there can be spill-over of fish and larvae or eggs from the reserve into adjacent fishing grounds.

Outside of the reserves an ecosystem-based approach requires the sustainable management of fisheries and other resources. It requires protection at the level of the whole ecosystem. This is radically different from the present situation where most fisheries management measures focus on single species and do not consider the role of species in the wider ecosystem. An ecosystem-based approach is also precautionary in nature which, in practice, means that a lack of knowledge does not excuse decision-makers from taking action, but rather they err on the side of caution. To achieve this, the burden of proof must be placed on those who want to undertake activities, such as fishing or coastal development, to show that these activities will not harm the marine environment, before the action is permitted. This will encourage sustainable development of fisheries, while limiting destructive practices.

As a way of implementing an ecosystem-based approach, Greenpeace is advocating the establishment of a global network of marine reserves covering 40% of the world's oceans. Because many marine ecosystems have become degraded and protection of biodiversity is vital, urgent action is needed to implement these marine reserves.

Foreword

I will never forget my first sight of a coral reef. I was 20 and Saudi Arabia my first foray abroad. The July heat in Jeddah topped 40°C as I pulled on mask and snorkel by a dusty roadside. Stretching ahead of me, lagoon waters shimmered inviting hues of green and brown. They darkened suddenly to indigo beyond a ragged line of breakers far offshore where the reef fell away into Red Sea depths. Hot lagoon brine gave scant relief from the scorching sun as I threaded my way among razor edged corals and fragments of storm-tossed rock. I struggled through a blizzard of surf at the reef crest until the water cleared beyond and the reef came alive around me. I had read widely on coral reefs that summer but none of it prepared me for the thrill as I entered a world beyond my imagination.

Fish surrounded me. Blue-sided surgeonfish, striped Abudefduf, clouds of orange Anthias hovering like butterflies, stately unicornfish, goggle-eyed porcupine fish, scowling eels. It was hard to take in even a tenth of what I saw. I felt like the Governor of a Newfoundland colony who wrote in 1620, "the Sea, so diversified with … Fishes abounding therein, the consideration of which is readie to swallow up and drowne my senses not being able to comprehend or expresse the riches thereof." Coral reefs have been around in one form or other for hundreds of millions of years and nature has had plenty of time to experiment. For variety of colour, form and sheer exuberance of life they are unmatched in the sea.

I had little reason to think then that life on this or other reefs would ever be different. The geological durability of rock forged by living coral over hundreds of thousands of years connected me with deep history. It gave the fleeting lives of the fish and invertebrates around me an unshakeable permanence. But I have come to realise that coral reefs have a tenuous hold on life today. Our own species could soon bring this ancient and remarkable ecosystem to an abrupt end. In the 25 years since my first encounter, coral reefs throughout the Indian and Pacific Oceans have lost an average of 1% of their coral cover per year. In the Caribbean, coral cover has declined from around 50% to 10% over the last 3 decades.

The many causes of coral loss are chronicled in this book and include overfishing, pollution, disease epidemics, development, global warming and ocean acidification, among others. They are problems that, to greater or lesser degrees, afflict every place, every habitat and every species in the sea. That the root causes of harm can all be traced to us is testament to the dominance humanity has now achieved over this planet. Even as little as 50 years ago, it was still possible to believe that much of the sea was beyond our reach and therefore wild. Monsters real and imagined lurked in unplumbed depths. Today they are mostly imagined because so many of the titans that once inspired fear and awe in seafarers have been lost to hunting and fishing.

While the challenge of recovering life in the sea is immense, there is much reason to hope. As this book reveals, sea creatures show astonishing resilience when we give them space to live lives free from harm. Where fully protected marine reserves have been established, stocks of food fish increase, habitats rebuild and productivity can be restored. The benefits continue to accrue for decades after protection commences and, like bank deposits, the longest established reserves deliver the greatest dividends.

With power comes responsibility. To date we have exercised much power but little responsibility in the sea. Our predecessors saw marine creatures as commodities and measured their worth in cash. Managers still focus too much on what we take from the sea, to the detriment of the many ways in which intact marine ecosystems benefit our lives. This book explains what we have done to the sea, what ocean life does for us, and why it is vital that we restore the richness and vitality of our oceans.

Callum Roberts
Environment Department
University of York, York
YO10 5DD, UK
Author of *The Unnatural History of the Sea*

Acknowledgements

With special thanks to: Lena Aahlby, Martin J. Attrill, Sonia Bejarano, Frida Bengtsson, Paloma Colmenarejo, Cat Dorey, Mark Everard, Truls Gulowsen, Mike Hagler, Alastair Harborne, Sara Holden, Lindsay Keenan, Sarah King, Oliver Knowles, Samuel Leiva, Sebastián Losada, Daniel Mittler, Giorgia Monti, Kate Morton, Femke Nagel, Karen Sack, Bettina Saier, Sarah Smith, Emma Stoner, Nina Thuellen, Reyes Tirado, Sari Tolvanen and Jim Wickens.

Contents

Chapter 1
Biodiversity

Abstract The term biodiversity refers to the total variability of life on earth. In the oceans about 300,000 species are known but it is likely that there are many more, undescribed species. This chapter discusses the biodiversity of a selection of marine habitats including the deep ocean, the open oceans, and, in the coastal zone, coral reefs, mangroves and seagrasses.

Keywords Biodiversity, deep oceans, seamounts, hydrothermal vents, coral reefs, mangroves, seagrasses, conservation

The deep oceans beyond the continental shelf regions are home to many species. For example, underwater mountains, known as seamounts, represent a treasure house of biodiversity. They support corals, sponges, anemones, rock-dwelling invertebrates and huge aggregations of many fish. However, the rich life on some seamounts has been destroyed by the highly destructive practice of bottom trawling, and stocks of some fish species have been severely depleted. To prevent further destruction it has been suggested by some that the United Nations should establish a moratorium on bottom trawling.

The deep oceans are also host to hydrothermal vents – geysers on the seafloor that gush hot water into the cold deep waters. Hydrothermal vents support a diverse array of very unusual life forms. Currently, one of the greatest threats to these life forms is bioprospecting, that is, the exploration of biodiversity for both scientific and commercial purposes. It is possible that mining the deep sea for metal ores may also threaten hydrothermal vent sites in the near future.

Out in the open oceans there are some features that enhance biodiversity. Where warm and cold currents mix, food becomes concentrated and this is a haven for wide-roaming animals such as tuna, sea turtles, seabirds and whales. Other features include upwellings where deep, cooler and usually nutrient-rich water moves to the ocean surface and supports many open-water fish. Drift algae, which floats on the sea surface, and is often associated with fronts and eddies, also provides vital support for many species at some stage in their life-cycle. All these features may ultimately need to be included in plans for marine reserves in order to protect the high biodiversity they support.

Adjacent to coastal regions and supporting high biodiversity are coral reefs, seagrass meadows and mangrove forests. These features are not only important in terms of their biodiversity but they also provide coastal protection from wave and storm impacts. However, all of these ecosystems are threatened by human activities. For example, commercial reef fisheries and overfishing has caused serious degradation of coral reefs and loss of biodiversity. Intense competition means that some fishers have resorted to the use of destructive fishing techniques including explosives and cyanide. Mangroves are threatened by the pressures of increasing populations, industrial and urban development and commercial aquaculture practices. An estimated 35% of the original area of mangroves has been lost in the past 2 decades. In regard to seagrass beds, there is increasing concern about worldwide losses due mainly to coastal development. Threats include dredging operations, nutrient pollution, industrial pollution and sediment inputs. Paramount to the protection of coral reef, mangrove and seagrass habitats will be the establishment of more and effective marine reserves. Moreover, because some species live in-between and use more than one of these habitats, it has been suggested that efforts should be made to protect connected corridors of these ecosystems.

1.1   Introduction to Biodiversity

The term biological diversity, or the shortened form, "biodiversity", refers to the total variability of life on earth. The United Nations Convention on Biological Diversity uses the following definition of biodiversity (UNEP 1995):

> *'Biological diversity' means the variability among living organisms from all sources, including, inter alia, terrestrial, marine and other aquatic ecosystems and the ecological complexes of which they are a part; this includes diversity within species, between species and of ecosystems.*

The most common usage of the term biodiversity refers to the number of species found in a given area, or species diversity (Gray 1997). For the earth as a whole, most estimates of the total number of species lie between 5 and 30 million. Of these, about two million species have been officially described, and the rest are unknown or unnamed. (Millennium Ecosystem Assessment 2005). More species are known of on land than in the sea. It has been estimated that there are, in total, approximately 300,000 known marine species (Gray 1997). The difference between land and sea is partly due to the very high diversity of beetles (Coleoptera) – 400,000 species are described. The difference may also be due to the fact that more study has been done on land than at sea (UNEP 1995).

At the level of phylum (the classification of all animals into broad groupings) marine diversity is close to twice that on land. For example, of the 33 animal phyla recognized in 1995, 32 occur in the sea; 15 are exclusively marine and 5 are nearly so (UNEP 1995). In contrast, only one phylum occurs exclusively on land. The most diverse ecosystems in the ocean, such as coral reefs, may have levels of species diversity roughly similar to the richest ecosystems on land, such as lowland tropical rain forests (UNEP 1995).

This chapter focuses on biodiversity of the deep oceans (Section 1.2), the open ocean pelagic zone (Section 1.3), coral reefs (Section 1.4), mangroves (Section 1.5) and seagrasses (Section 1.6).

1.2 The Deep Oceans

The deep sea covers about 70% of the Earth and has an average depth of 3,200 m (Prieur 1997). Of the oceans that lie beyond the continental shelves, 88% are deeper than 1 km and 76% have depths of 3–6 km (UNEP 2006a). Despite the darkness of the deep ocean, near freezing temperatures and scarce energetic supplies, the deep ocean supports a surprisingly high diversity of species. This was first discovered by sampling of deep sea sediments in the 1960s by the Woods Hole Oceanographic Institute (Grassle 1991).

About 50% of the deep ocean floor is an abyssal plain comprised mainly of mud flats. Superimposed on the deep sea bed are other deep sea features including submarine canyons, oceanic trenches, hydrothermal vents and underwater mountains called seamounts. This great variety of benthic (sea bottom) habitats support multitudinous life forms (UNEP 2006a). On the sea bed, the large animals (megafauna) are dominated by echinoderms (including brittle stars, sea stars and sea urchins). Other species present are crustaceans (including crabs), giant sized crustaceans called amphipods, sea cucumbers, bristle worms, sea spiders, bottom dwelling fish, sponges, benthic jellyfish, deep sea corals, deep sea barnacles, sea squirts and bryozoa (moss animals) (Gage and Tyler 2001). Koslow et al. (1997) noted that about 2,650 species of demersal (bottom dwelling) deep sea fish are known and suggested that, in total, there are probably about 3,000–4,000 species.

In addition to the marine animals listed above which dwell on or near to the seabed, a very high diversity of small animals are also found within the sediments of the deep sea. These animals are classified according to their size, the largest of which, macrofauna, are those retained by sieves with meshes of about 1 mm while even smaller animals, meiofauna, are retained by the finest screens down to a mesh opening of 62 μm or smaller. Together, the macrofauna and meiofauna comprise the most diverse component of the deep sea benthos (Gage and Tyler 2001). For instance, one study which sampled about 50 m² of sediments, identified 707 species of polychaete worms and 426 species of peracarid crustaceans (Grassle 1991). Other species of macrofauna which have been found in deep sea sediments are different sorts of worms, mites, other crustaceans, amphipods and molluscs. Meiofauna include nematode worms, certain copepods and crustaceans and single-celled organisms known as Foraminifera (Gage and Tyler 2001).

It was suggested in the early 1990s that the number of undescribed species in the deep sea may be as high as ten million. However, this was contested by others at the time and a more realistic suggestion of 500,000 was proposed (Gray 1997).

Deep water fisheries, that is fishing at depths of over 500 m, has increased in recent years as traditional fisheries on continental shelves have declined. However, the practice of deep water fishing is known to be extremely destructive to deep water marine ecosystems. Deep water fish species are often long-lived and slow to reach maturity and, hence, breeding potential, which renders them especially vulnerable to exploitation. Further detail on the biodiversity of seamounts and the need to conserve and protect them from deep sea fishing is discussed below.

1.2.1 Seamounts

Seamounts are undersea mountains which rise at least 1,000 m above the sea floor. Smaller undersea features with a height of between 500 and 1,000 m are defined as knolls and those rising less than 500 m high are defined as hills. The term seamount is, however, often also used to refer collectively to seamounts, knolls and hills.

Seamounts are usually formed by volcanic activity and they often occur in chains or clusters. Using the classification of seamounts being over 1,000 m high, it has been estimated that there are over 30,000 seamounts in the Pacific ocean, about 810 in the Atlantic and an indeterminate number in the Indian ocean (Rogers 1994). Kitchingman and Lai (2004) suggested that there may be over 50,000 seamounts worldwide. Plate 1.1 shows the location of known seamounts on a world map.

Despite some 30 years of studies directed at seamounts, the knowledge base on biodiversity remains poor (Johnston and Santillo 2004). For example, of the many thousands of seamounts, animal life has only been studied on almost 200 (Stocks 2004a). However, research to date on seamounts shows that these unique marine habitats represent a treasure house of biological diversity (McGarvin 2005). For example, according to Stocks (2004b), the total number of species found to be associated with seamounts to date was around 2,700. There is generally also a high abundance of life on seamounts, which has led to descriptions of them as 'underwater oases' (Stocks 2004b). Life is abundant at all trophic levels, from tiny plankton to rock-dwelling invertebrates and aggregates of fish populations in their vicinity (Fock et al. 2002; Koslow et al. 2001; Rogers 1994).

Water currents are enhanced around seamounts. Due to these enhanced currents, the animal life forms on the seamounts are dominated by suspension feeders (animals that feed by straining suspended matter and food particles from water) (Koslow et al. 2001). These animals include corals, anemones, featherstars and sponges (Stocks 2004b). They make seamounts visually striking and consequently seamounts have been likened to underwater gardens due to the branching tree-like and flower-like corals and sponges that cover many of them. Some stands of corals which have been discovered on seamounts are several centuries old. In addition to reef-building organisms such as corals and sponges, other invertebrate species common to seamounts are crustaceans, molluscs, sea urchins, brittle stars and polychaetes (bristle worms) (Stocks 2004b).

Many species of fish are associated with seamounts (Rogers 1994) and some are well known for the huge aggregations they form over these features. A study cited by Tracey et al. (2004) described 263 species of fish that were found on seamounts in the New Caledonian region. Perhaps the best-known fish that are associated

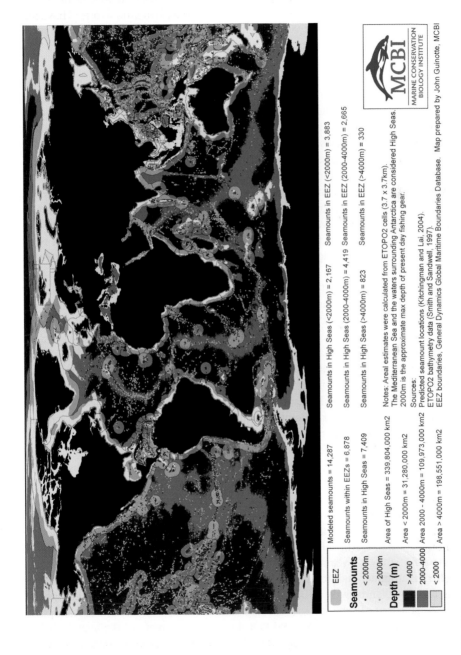

Plate 1.1 Global map of known seamounts

with seamounts are those that have been commercially exploited. These include the orange roughy (*Hoplostethus atlanticus*) which has a global distribution, alfonsinos (*Beryx decadactylus*) in the tropics and subtropics and the pelagic armourhead (*Pseudopentaceros richardsoni*) of the southern Emperor and North Hawaiian Ridge seamounts (McGarvin 2005; Rogers 1994). Migratory species such as tuna, marine mammals and seabirds are known to congregate over seamounts, which suggests they are important for these species (Stocks 2004b).

New species of marine organisms have been found on almost every seamount which has been studied. For instance, one study of seamounts off southern Tasmania found that 24–43% of the invertebrate species collected were new to science (Koslow et al. 2001). It is likely that most unsampled seamounts will also include as yet undiscovered species (Stocks 2004b). On some of the seamounts which have been studied, a high number of species that are endemic have been reported, that is, species which are restricted to only one seamount or seamount chain and have not been found elsewhere in the oceans. Stocks (2004b) cited studies which showed high levels of endemism; for example, 31–36% of species found in seamounts south of New Caledonia were endemic, and in the Pacific off the coast of Chile on two seamount chains, 44% of fishes and 52% of bottom-dwelling invertebrates were endemic. Because knowledge of all species in the oceans is incomplete, it is said that seamounts have 'apparently' high rates of endemism as it is not yet possible to know whether the species present do occur elsewhere in the oceans (Stocks 2004b).

Seamounts have come under intensive exploitation by trawl fisheries since the 1960s. The development of seamount fisheries has taken place in relative ignorance of the biology and ecology of seamounts (Johnston and Santillo 2004). For example, the life history of most seamount species is unknown but research on some species has shown they are generally long-lived (of the order of about seventy to hundreds of years) and slow growing (Koslow et al. 2001). This means that they are vulnerable to depletion by fishing because stocks are not replenished quickly. In addition, where species are endemic to just a single seamount or area of seamounts they rapidly become vulnerable to extinction when exploited (Stocks 2004b).

Fishing on seamounts has led to the depletion of some species and causes damage to communities of corals and other bottom dwelling life-forms (see also Chapter 2, Section 2.5.1.1 on bottom trawling). Stocks of pelagic armourhead (*Pseudopentaceros richardsoni*) over Pacific seamounts northwest of Hawaii were depleted to the point of commercial extinction in less than 20 years. Stocks of orange roughy (*Hoplostethus atlanticus*) have been severely depleted on some seamounts around Australia and New Zealand (Johnston and Santillo 2004). A scientific study of seamounts off the coast of southern Tasmania found that unfished seamounts had 46% more species per sample taken from the surface of the seamount than heavily fished seamounts. In addition, the biomass (weight of living organisms) of samples was 106% greater from unfished seamounts (Koslow et al. 2001). The study also reported that the impact on reefs of the heavily fished seamounts was dramatic, with the coral substrate and associated community largely removed. Other studies on seamount trawling have also shown that coral pieces are a common by-catch of fishing on seamounts. Damage to corals, sponges, anemones and other reef organisms on seamounts is particularly of concern because these species provide habitat for many other organisms (Stocks 2004b).

In summary, seamounts are unique habitats of the deep oceans that support multitudinous species in abundance, some of which are endemic. Seamounts also appear to be important to other migratory species. The fishing practice of bottom-trawling on seamounts is highly destructive to these rich havens of marine life. Greenpeace is campaigning for the UN to impose a moratorium on the practice of bottom-trawling (see also Chapter 2, Section 2.5.1.4). Such a moratorium would provide for the widespread protection of these fragile environments and make it possible to undertake the full scientific assessments needed to develop the permanent solutions necessary to conserve these rich and vulnerable deep sea ecosystems. Key to their long-term protection is the establishment of no-take Marine Reserves. Relatively few seamounts have so far been designated as Marine Reserves or Marine Protected Areas (Johnston and Santillo 2004). Greenpeace is campaigning for the implementation of a global network of Marine Reserves which would protect at least 40% of the world's oceans including particularly vulnerable areas such as seamounts from fishing and other threats such as seabed mining (Roberts et al. 2006) (see Chapter 7).

1.2.2 Deep Sea Hydrothermal Vents

A hydrothermal vent is a geyser on the seafloor which gushes hot water into the cold, deep ocean. These hot springs are heated by molten rock below the seabed (Gage and Tyler 1991). Deep-sea vents have been found in the Pacific, Atlantic, Arctic and Indian Oceans (Little and Vrijenhoek 2003). They are primarily concentrated along the earth's Mid-Oceanic Ridge, a continuous underwater mountain chain that bisects the oceans and is a 60,000 km seam of geological activity. It is thought that hundreds, if not thousands of hydrothermal vent sites may exist along the Mid-Oceanic Ridge but as yet only about 100 sites have been identified because they are very hard to find (Glowka 2003).

In 1977, scientists discovered that vents were populated with an extraordinary array of animal life. This was most unexpected because the environment around vents appears to be hostile to animal life – the fluid coming from vents is hot (up to 390°C), anoxic (without oxygen), is often very acidic and is enriched with hydrogen sulphide, methane and various metals (particularly iron, zinc, copper and manganese) (Little and Vrijenhoek 2003). However, these seemingly toxic fluids are now known to support high densities of animal communities.

After 30 years of research at hydrothermal vent sites, a total of over 550 species have been described (Ramirez-Llodra et al. 2007). The animals at vent sites are unique in that they do not rely ultimately on sunlight as an energy source for food, but on chemosynthetic bacteria which live off hydrogen sulphide in the vent fluids. Animals inhabiting vent sites survive either by consuming the free-living bacteria directly as food or by having a symbiotic relationship with them in which bacteria live within their tissues (Gage and Tyler 1991). It has been reported that vent environments support one of the highest levels of microbial diversity on the planet (Glowka 2003).

Animal species inhabiting vent sites include molluscs, gastropods, tube-dwelling worms, sea anemones and crustaceans (see for example Plate 1.2) (Gage and Tyler 1991; Little and Vrijenhoek 2003). Some species of fish have been found to live within vent environments and more live in their vicinity (Biscoito et al. 2002). Many animal species are exclusively native (endemic) to vent sites (Little and Vrijenhoek 2003). There is variation in the species existing at different vent sites (Van Dover 2005). Although the number of species found at any one vent site is relatively low (low species biodiversity), the abundance of animals is generally high. For example, enormous densities of a giant clam-like organism (*Calyptogena magnificica*) and a giant mussel (*Bathymodiolus thermophilus*) have been found in the area of vents of the Eastern Pacific (Gage and Tyler 1991). Typically, most of the species diversity at vent sites is attributed to small inconspicuous animals, but sites are dominated by a few large and visually striking species such as vestimentiferan tube worms (Siboglinidae), vent clams (Vesicomyidae) and mussels (Bathymodiolinae) and the blind vent shrimp (*Rimicaris exoculata*) (Little and Vrijenhoek 2003).

Plate 1.2 A dense bed of hydrothermal mussels covers the slope of the Northwest Eifuku volcano near the seafloor hot spring called Champagne vent. Other vent animals living among the mussels include shrimp, limpets, and Galatheid crabs (Pacific Ring of Fire 2004 Expedition. NOAA Office of Ocean Exploration; Dr. Bob Embley, NOAA PMEL, Chief Scientist)

1.2.2.1 Protection

The more accessible hydrothermal vents are potentially threatened by a number of human activities such as seabed mining for polymetallic sulphide deposits, submarine-based tourism and marine scientific research (Glowka 2003). Glowka proposed that of these, marine scientific research poses the greatest threat for the most visited vent sites due to, for example, concentrated sampling practices. It was suggested therefore that marine scientific research needs to be placed on a more sustainable footing at hydrothermal vents. Indeed, the unique biodiversity of deep-sea hydrothermal vents needs to be protected from all potentially human-based destructive practices. The most appropriate tool for protecting these environments is their designation as fully protected marine reserves. In this regard, there has been some headway in initiating protection for a few sites, namely the Endeavour Hydrothermal Vents Area (located in Canadian waters on the Juan de Fuca Ridge about 256 km southwest of Vancouver Island) and Lucky Strike and Menez Gwen vent fields in the North-east Atlantic Ocean (within Portugal's EEZ). For Endeavour, a proposal was made in 2002 for it to become a Marine Protected Area and this would mean that nothing could be removed from the area without a relevant license and submission of a research plan (Glowka 2003). In March 2003, the Endeavour Marine Protected Area was legalised by the Canadian government under the Oceans Act (Fisheries and Oceans Canada 2006). For Lucky Strike and Menez Gwen, the Worldwide Fund for Nature (WWF) worked with the Azores regional government to have these relatively shallow vents designated as Marine Protected Areas in 2002 (WWF 2008).

1.2.3 Bioprospecting in the Deep Sea

Bioprospecting is the exploration of biodiversity for both scientific and commercial purposes. There is no internationally accepted definition of bioprospecting. However, it can be summarised as the investigation of an area's biodiversity, and sampling of biological organisms for scientific research or commercial purposes. It is often difficult to differentiate between research on genetic resources for purely scientific purposes, and that for commercial activities. Generally, such resources are collected and analysed as part of a scientific research project, often as partnerships with scientific institutions and industry. At a later stage these resources and findings can enter the commercial arena as products derived from information discovered during these scientific endeavours (Greenpeace 2005a).

With advances in technology, bioprospecting has started to take off in the marine context. There is a realisation that many marine plants, animals and microorganisms contain unique biochemicals which could be integral to developing new products for use in the health, pharmacology, environmental and chemicals sectors.

Compounds that have been isolated from various marine species such as sponges, corals and sea slugs are now being sold commercially (Greenpeace 2005 f).

To date, most bioprospecting in the marine environment has taken place in shallower waters. However, scientists are beginning to appreciate the valuable resources that are housed in the depths of the high seas. Many of these deep-sea species have developed unique biological and physiological properties in which to survive in these extreme environments. These include slow growth, late sexual maturity, the ability to withstand cold, dark and highly pressurised environments, as well as a high level of endemism in many ecosystems. It is these properties that are attracting the interest of the scientific and commercial sectors. Yet these same properties make deep-sea species highly susceptible to disturbance and change (Greenpeace 2005 f).

At the moment, bioprospecting in the deep sea is still restricted to a very select sector – either commercially based or academic – that has both the technical and financial capital to exploit these resources. However, with ever-developing advances in technology, the opportunity to exploit these little-known resources is increasing. Continued scientific research to extend our knowledge of deep-sea ecosystems is important, but there are potentially detrimental impacts of this exploration, including physical disturbance or disruption of ecosystems, pollution and contamination, as well as problems of over-harvesting (Greenpeace 2005 f).

A potential future threat to deep-sea ecosystems from seabed disturbance is deep-sea mining for metal ores. For instance, it has recently been announced that a specialised deep-sea submersible is being developed for intended future use in dredging the seafloor for copper, gold and zinc (Heilman 2006). The submersible will be capable of reaching depths of 1,700 m and is scheduled to be ready for use in 2009. Mining is likely to happen around hydrothermal vents because such areas can contain high levels of mineral sediments. Disturbance by mining operations is a threat to the biodiversity of these regions.

Since much of the deep oceans lie beyond national jurisdiction there is currently no legal regime to regulate bioprospecting in the deep oceans. As such, it poses a threat to deep-sea ecosystems, which due to their unique biological characteristics are particularly vulnerable to habitat disturbance. Greenpeace believes that bioprospecting must be managed in a way that would minimise the potential impact and disruption to deep-sea ecosystems. This requires clear international regulations, which currently do not exist (see also Chapter 6, Section 6.3). What is needed is an integrated, precautionary and ecosystem-based management approach to promote the conservation and sustainable management of the marine environment in areas beyond national jurisdiction, including equitable access and benefit sharing of these resources. This could be provided by a new implementing agreement under the United Nations Convention on the Law of the Sea (UNCLOS). Greenpeace has suggested the necessary prerequisites for a new agreement in a report "Bioprospecting in the Deep Sea" (Greenpeace 2005 f). Only concerted international action by States to put such a legal framework into place will ensure the conservation and sustainable management of the planet's final frontier – the high seas.

1.3 Biodiversity Hotspots at Sea in the Pelagic (Open Water) Zone

The abundance and diversity of biological communities across the open oceans is only just beginning to be understood (Malakoff 2004; Worm et al. 2005). It has been found that some features of the pelagic zone are particularly favourable to enhancing biodiversity. These include areas where warm and cold currents mix. For example, wide-roaming pelagic animals such as tuna, sea turtles, seabirds and whales are known to follow "oceanic fronts", where cold and warm water masses collide, and to congregate in other areas where food is concentrated. Recently, an area of about 125,000 km² off the Baja coast of California was identified where the cool southbound California current collides with a warmer northbound stream. Fishing records showed that this area had supported very high landings of sword-fish and striped marlin over the past 35 years and other research showed that blue whales tended to linger in the region (Malakoff 2004).

Other features that are important for many pelagic fish are upwellings where deep and dense, cooler and usually nutrient-rich water moves towards the ocean surface and replaces the warmer, usually nutrient-deficient, surface water. The nutrients may then be used to support the growth of phytoplankton. In turn, phyto-plankton may feed zooplankton, which are important in the food-chain for pelagic fishes. Upwelling systems are known to be important to many pelagic fishes and sustain a large proportion of the world's fisheries (Paramo et al. 2003; Pauly and Christensen 1995; Ward et al. 2006).

A recent study investigated the global diversity of two predatory fish species, tuna and billfish, in the open oceans (Worm et al. 2005). The study revealed that there were peaks of diversity at intermediate latitudes (15°–30° North or South) and lower diversity towards the poles and the equator. The same latitudinal distribution was also found for zooplankton diversity. These results suggested that this global pattern of diversity could be general across several trophic levels (i.e. levels of the food chain from tiny zooplankton animals to fish that prey on the zooplankton to predatory fishes). The study also showed that the species diversity of zooplankton, tuna and billfish was linked closely to sea surface temperature and oxygen con-centration, such that optimal habitats were characterised by warm waters (about 25°C) with sufficient oxygen concentrations. Fronts and eddies were also found to be important in supporting high diversity due to the concentration of food in these areas. The protection of such areas of high biological diversity in the pelagic zone is discussed below (Section 1.3.1).

Another pelagic habitat of great importance to marine biodiversity is drift algae. Drift algae accumulations float on the sea surface in some areas of the open oceans, existing as occasional clumps, expansive mats of up to several kilometres, or elongated lines. It provides vital support for many species at some stage in their lifecycle. This includes at least 280 species of fish, many invertebrates, 4 species of turtle in early life and some seabird species. Drift algae is thus an important habitat to many species in the open oceans. However, in some areas it is under threat from commercial harvesting for food, livestock fodder, fertiliser and medicine. Other threats to drift algae include commercial fishing, which is in direct association with algal mats, vessel traffic through drift algae habitat, and pollution (Hemphill 2005).

Paramount to the protection of open water features such as drift algae, upwellings and oceanic fronts is the establishment of marine reserves (see Chapter 7, Section 7.4.3).

1.4   Coral Reefs

Coral reefs are distributed in shallow seas throughout the tropics (Fig. 1.1). About one third of tropical coastlines are made of coral reefs and they occur in over 100 countries. Globally, coral reefs are estimated to occupy 284,300 km^2, an area which represents less than 1.2% of the world's continental shelf area. There are two distinctive regions of coral reefs worldwide, one centred around the wider Caribbean (the Atlantic), and the other reaching from East Africa and the Red Sea to the Central Pacific (the Indo-Pacific) (Spalding et al. 2001).

The majority of corals require a solid surface on which to grow, and this limits their formation to rocky substrates in the tropics. They cannot grow on fine muds or mobile sediments, so they are largely absent from river mouth areas and along stretches of sediment-laden coastlines (Spalding et al. 2001). Living coral is only a thin veneer, measured in millimetres, often covering the thick limestone structures (reefs) that the corals have laid down over very long periods. Archipelagoes consisting of hundreds of atolls, such as the Marshall Islands and the Maldives, are formed from coral reefs. The Great Barrier Reef in Australia is over 2,000 km long (Birkeland 1997).

Reef corals build calcium carbonate (limestone) skeletons and most derive at least some of their nutrition from photosynthesis by algae which live within their tissues (Sebens 1994). The ability of corals to construct massive calcium carbonate frameworks sets them apart from all other marine ecosystems (UNEP 1995). The resulting reef structure is complex and contains a superabundance of surfaces and spaces, many of them internal (Adey et al. 2000). This provides habitat and shelter for numerous species. For example, at the times of night or day that fish are inactive, many retreat into the reef to take shelter (Sale 1991).

1.4.1   Biodiversity of Coral Reefs

Coral reefs are the most biologically diverse ecosystems of the entire oceans. Their high species diversity has led to them being called "rainforests of the sea". As many as 100,000 reef-dwelling species may have been named and described to date (Spalding et al. 2001) although estimates of the total number of species

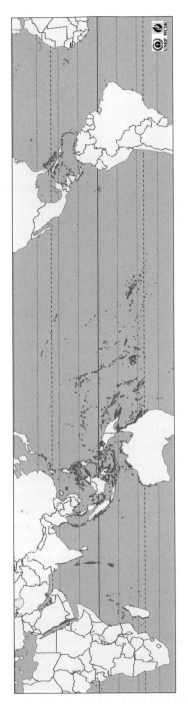

Fig. 1.1 World distribution of scleractinian corals (Spalding et al. 2001a, reproduced with permission from UNEP)

on the world's coral reefs of both one million and three million species have been proposed (Adey et al. 2000). The species diversity of corals and reef-dwelling organisms differs geographically, such that diversity declines going northwards and southwards of the equator (Sebens 1994). However, at a global scale, reefs appear to support high diversities of fish regardless of latitude (Ormond and Roberts 1997). Centres of particularly high biodiversity of coral reefs are located in the tropical Indo-West Pacific Ocean, called the East Indies Triangle (Briggs 2005).

Reefs in the Atlantic have a lower diversity of species than Indo-Pacific reefs. In the most biologically diverse coral reefs centred in the Philippines, Indonesia, Malaysia and Papua New Guinea there are 500–600 species of coral (Spalding et al. 2001). There are many other species inhabiting coral reefs including sponges, hydrozoa, jellyfish, anthozoans, worm-like animals, crustaceans, molluscs, echinoderms (such as starfish and sea urchins), sea cucumbers and tunicates (Spalding et al. 2001). It has been estimated that there are at least 4,000, perhaps even 4,500 species of marine fishes inhabiting the world's coral reefs. This represents more than a quarter of all marine fish species (Ormond and Roberts 1997). Sea turtles often make use of reefs as a source of food, and a number of seabird species are found regularly in coral reef environments. Several species of marine mammals can also be found in close proximity to coral reefs; for instance, dolphins regularly take shelter in bays and lagoons near reefs and sometimes feed on reef animals. Humpback whales breed close to coral reefs in Hawaii, the Great Barrier Reef and the Caribbean (Spalding et al. 2001).

Some coral reef organisms are also dependent on neighbouring seagrass beds and mangroves. For example, some coral reef fish use seagrass beds and mangroves as nursery grounds and some herbivorous fish forage in seagrass habitats by day and shelter on the reefs at night (Moberg and Folke 1999; Valentine and Heck 2005).

New species are still being discovered on coral reefs: in 2006, research by Conservation International in a region known as the Bird's Head Seascape off the coast of Indonesia's Papua Province, found more than 50 new species (Conservation International 2008; ENN 2006). This included 24 fish species and 20 species of coral that are new or likely to be new to science. Among the new fish discovered were two species of bottom-dwelling sharks, which use their pectoral fins to 'walk' across the seafloor. Biodiversity in the area is very high and Conservation International are now helping to establish a regional network of community and government-endorsed marine protected areas.

Figure 1.2 shows a diagrammatic representation of a healthy coral reef. A healthy reef contains species that perform critical functions for the maintenance and survival of the reef. For instance, herbivorous fishes can be divided into three different functional groups, namely scrapers, grazers and bioeroders. Scrapers remove algae and sediment from the reef and facilitate the settlement and growth of new corals. Grazers remove seaweed and thereby reduce coral shading and overgrowth by such species (Bellwood et al. 2004). Without herbivores the reef would be overgrown by faster-growing algae (Moberg and Folke 1999). Bioeroding fishes and urchins remove dead corals and expose the hard reef matrix for settlement by new corals and coralline algae (Bellwood et al. 2004). Besides herbivorous species, there

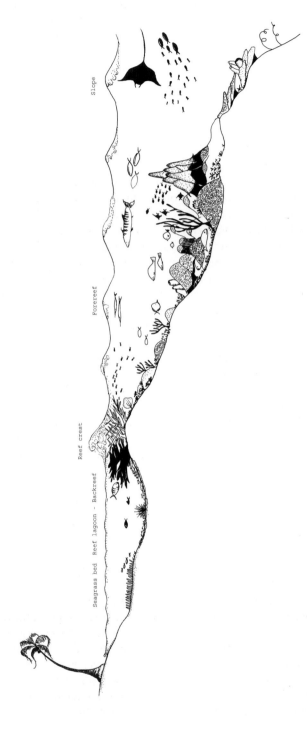

Fig. 1.2 Diagrammatic representation showing a profile through a typical healthy coral reef from slope to lagoon, including associated seagrass beds and sandy beach (Prepared by Sonja Bejarano)

are other species which perform particular functions on coral reefs. For example, predatory fishes regulate the number of herbivorous fishes and sea urchins (Moberg and Folke 1999). There is a danger that over-exploitation of coral reefs by fisheries can deplete species which have a similar function. Inevitably, this leads to the loss of that function on the reef and has detrimental consequences. Overfishing and the consequences of removing functional groups are discussed in Section 1.4.4.1.

1.4.2   Protection of Neighbouring Environments

Aside from their role as habitat for numerous marine species, coral reefs also provide an important role in shoreline protection. In this regard, coral reefs act as breakwaters, serving to protect beaches and coastlines from storm surges and wave action (Birkeland 1997; UNEP-WCMC 2006). Coastline communities that do not have coral reefs suffer greater damage during typhoons than those sheltered by coral reefs (Birkeland 1997). The tsunami of 26 December 2004, a devastating event, caused massive damages and incalculable suffering to millions of people around the Indian Ocean. There is anecdotal evidence and satellite photography, before and after the tsunami event, which suggests that coral reefs, mangrove forests and other coastal vegetation provided protection from the impacts of the tsunami. In Sri Lanka, some of the most severe damage occurred along coastlines where mining and damage to coral reefs was heavy in the past (UNEP 2006b).

The sheltering capacity of coral reefs physically creates favourable conditions for the growth of seagrass beds and mangroves (Moberg and Folke 1999). Coral reefs are also responsible for generating fine white sand which supplies tropical coastlines and is one of the main attractions in beach tourism (Moberg and Folke 1999).

1.4.3   Human Use of Coral Reefs

Fisheries: Coral reefs are harvested for a variety of seafood products such as fish, mussels, crustaceans, sea cucumbers and seaweeds (Moberg and Folke 1999). In some regions, large numbers of species are taken for consumption. For example, 209 species are taken at Bolinao, Philippines, about 250 species in the Tigak Islands, Papua New Guinea, and about 300 species around Guam (Birkeland 1997).

Coral reef fisheries provide food and livelihood for tens of millions of people throughout tropical and subtropical regions (McManus et al. 2000). It is estimated that there are 30 million small-scale fishers in the developing world and most are dependent to some extent on coral reefs for their livelihood (UNEP-WCMC 2006). Reef-associated fisheries are estimated to make up about 10% of total world marine fishery landings, according to United Nations Food and Agricultural Organisation statistics (Sadovy 2005).

Presently, reefs not only support subsistence fishers but also commercial reef fisheries, which trade products on the global market. Commercial reef fisheries supply export markets around the world as well as the restaurant and hotel industries. There is a live fish trade in South-East Asia that is supplied with coral reef products from the Pacific and Indian Oceans (UNEP-WCMC 2006). Fishing pressure on many reefs is not sustainable and overfishing has caused serious degradation of reefs and loss of biodiversity (see Section 1.4.4.1).

Tourism: Coral reefs support recreational activities including scuba diving (Plate 1.3), snorkeling and glass-bottom boat operations (UNEP-WCMC 2006). The income generated from such tourist activities is enormous (Moberg and Folke 1999). Scuba diving provides the main base of economies in a number of tropical developing countries and also contributes significantly to tourism revenue in more developed countries such as Australia and the USA (Birkeland 1997).

Pharmaceuticals: Marine organisms may contain compounds that are of use as pharmaceuticals. Coral reef organisms have provided an HIV treatment and a painkiller. In addition, some cancer drug research is investigating coral reef species (UNEP-WCMC 2006).

Others: Some coral reef species are collected for use as jewellery or souvenirs. These provide significant export revenue. However, the souvenir trade is largely unregulated (UNEP-WCMC 2006).

The marine aquarium trade is responsible for the collection and sale of many live corals, fish and invertebrates from coral reefs. These are sold mainly to consumers

Plate 1.3   Gotta Abu Ramad reef, Egypt (Greenpeace)

in the USA and Europe and, to a lesser extent, Japan (Wabnitz et al. 2003). It is of concern that the capture of live fish for the marine aquarium trade often involves the use of cyanide to stun creatures and then capture them (Moberg and Folke 1999). The use of cyanide can result in poisoning and killing of other non-target animals, including corals. In addition, the over-exploitation of some species and high levels of mortality associated with inadequate handling and transport of living organisms undermine the potential for a sustainable trade. The collection of organisms for marine aquaria occurs mainly from reefs in Southeast Asia but also increasingly from several island nations in the Indian and Pacific oceans. Only a small number of countries have comprehensive regulations to control the collection of marine ornamental species and it has been suggested that appropriate management is needed (Wabnitz et al. 2003).

Mining of live and dead coral for building materials has caused extensive degradation of reefs in a number of countries including The Maldives, Indonesia, Sri Lanka, Tanzania and the Philippines.

Impacts on reefs from mining have included declines in coral cover, diversity and fishes. Reefs that were mined before the mid-1970s have shown little recovery (Brown 1997).

1.4.4   Threats to Coral Reefs

It is of great concern that coral reefs are in decline globally (Pandolfi et al. 2005). Problems include a loss in coral cover and biodiversity coupled in some areas, such as the Caribbean and southern Florida, with a shift towards fleshy seaweed (macro-algae) dominated ecosystems (Bellwood et al. 2004; Szmant 2002). A report on the status of the world's coral reefs indicated that there is a continuing decline in reefs that are adjacent to large human populations in Eastern Africa, South, Southeast and East Asia, and the Caribbean (Australian Institute of Marine Science 2004).

It has been estimated that about 20% of the world's coral reefs have been destroyed and show no immediate signs of recovery; a further 24% are under imminent risk of collapse through human pressures; and another 26% are under longer-term threat of collapse (Australian Institute of Marine Science 2004). The principal threats to reefs are overfishing and nutrient and sediment pollution from poor land management practices. More recently, mortality of corals has increased due to increases in the prevalence of coral diseases and the emergence of coral bleaching due to increases in sea temperature (Australian Institute of Marine Science 2004; Pandolfi et al. 2003).

1.4.4.1   Overfishing

Historically, records show that coral reef ecosystems suffered from degradation even before 1900 and that this was likely due mainly to overfishing (Pandolfi et al.

2003). In more recent times, human populations have been expanding throughout large areas of the developing world which has led to increased fishing pressure on coral reef ecosystems (Roberts 1995). Moreover, commercial fisheries have entered the realm of reef fishing. The appearance of new, non-traditional fishers on reefs has led to intense competition and the use of destructive fishing implements such as explosives and poisons (Pauly et al. 2002).

Most coral reefs that are in the range of small fishing boats are now overfished (Australian Institute of Marine Science 2004). A global survey of over 300 reefs in 31 countries reported that overfishing had occurred on most of the reefs and key species of fish and invertebrates that were surveyed were reduced to low levels (Hodgson 1999). Increasing human pressure on a coral reef inevitably leads to a situation where more fishers chase progressively fewer fish. For example, firstly, catches of large target species tend to decline. Then fishers target all fish species using more efficient methods of traps, fine mesh nets and spears. Finally, explosives and poisons are used to catch the few remaining fish (Australian Institute of Marine Science 2004; Roberts 1995).

Overfishing on reefs leads to a reduction in species diversity. All but the smallest fishes may be removed (Roberts 1995). For example, there are many reefs in Eastern Africa, South and Southeast Asia and the Caribbean where it is rare to see a fish over 10 cm long (Australian Institute of Marine Science 2004). About 50 coral reef fishes are listed as Threatened by the IUCN World Conservation Union and for most of these species, this is due to exploitation (Sadovy 2005).

Overfishing is often accompanied by destructive fishing methods, that is, the use of explosives and the use of poisons, usually cyanide. These methods of fishing are particularly harmful to coral reef ecosystems. Fishing with explosives is largely restricted to Southeast and East Asia and is usually applied when fish stocks have become depleted, making traditional fishing methods unprofitable (Australian Institute of Marine Science 2004). A dynamite blast kills all fish within a 50–70 m radius. A proportion of the dead fish float to the surface where they are collected, but the rest sink to the seafloor and are left. As well as destroying fish, fishing with explosives damages large areas of coral, resulting in decreased coral cover (Guard and Masaiganah 1997).

Fishing with cyanide was first employed by the marine aquarium trade but is now also used to capture live fish for restaurants, particularly the growing demand for live reef fish for the Chinese restaurant trade. The use of cyanide leads to serial depletion of large coral reef fishes and also results in the death of other coral reef species (Australian Institute of Marine Science 2004). Divers frequently destroy live corals around holes and crevices to collect stunned fish (Halim 2002). Fishing with both explosives and cyanide destroys fish habitats and therefore by definition these methods are unsustainable (Australian Institute of Marine Science 2004).

As discussed in Section 1.4.1 above, species on coral reefs may be put into groups according to their function. Overfishing a broad array of species may remove functionally similar species, leading to loss of that function from the ecosystem (Roberts 1995). For example, in places in the Caribbean such as Jamaica, depletion of herbivorous fish by overfishing left only sea urchins as grazers to control the growth of seaweed (macro-algae). In the early 1980s, a disease struck the sea urchin population and brown fleshy seaweed rapidly overgrew the reefs in some places. Although sea urchins have since recovered their numbers, the algae remain dominant on the affected reefs (Scheffer et al. 2005). Studies on the decline of reefs to algal dominance have also determined that a reduction of the abundance of herbivorous fishes is an important factor in the overgrowth of reefs by macro-algae (McManus et al. 2000; Thacker et al. 2001). At Glovers Reef in Belize, fishing of herbivores is considered to be an important factor in the reduction of coral by 75% and the increase in macro-algae by 315% over a 25-year period (McManus et al. 2000).

It is possible that overfishing and loss of a functional group is also responsible for the massive outbreaks of crown-of-thorns starfish on the Great Barrier Reef, Australia, since the 1960s. The outbreaks have resulted in the considerable reduction in coral cover (Brown 1997). A possible explanation for the outbreaks is overfishing of the species that prey upon the larval or juvenile stages of the crown-of-thorns starfish (Jackson et al. 2001a). The outbreaks may also be due to increased nutrient pollution supporting an abundant food supply for the larval starfish (Brown 1997).

1.4.4.2 Land-Based Sources of Pollution

Intensified land use and urbanisation often increase run-off of pollutants, nutrients and sediment particles and cause major problems for coral reefs (Moberg and Folke 1999).

Sedimentation: Developments in coastal areas can increase the flow of sediments onto coral reefs (Australian Institute of Marine Science 2004). When sediment is stirred up in the water, light penetration is reduced and this may affect photosynthesis on coral reefs. Sediments may also smother corals. Such effects from increased sedimentation can result in coral mortality. For example, at Ko-Phuket, Thailand, dredging for a deep-water port over an 8-month period caused a significant decrease in coral cover on reefs adjacent to the activity. A year after the dredging, the corals were showing rapid signs of recovery (Brown 1997).

Sediment release into the oceans is increasing due to the development of coastal areas for expanding populations and increases in agriculture (Australian Institute of Marine Science 2004). Aquaculture operations can be a considerable source of both sediments and nutrients into coastal waters (Edinger et al. 1998). Clear felling of tropical forests in coastal zones can also be a major contributor to increased sedimentation (Australian Institute of Marine Science 2004).

Nutrient Pollution: Nutrient releases to the oceans, especially of inorganic and organic nitrogen and phosphorous compounds, have increased dramatically in recent decades. Sources of nutrients include agricultural run-off, sewage and aquaculture waste. Excessive nutrients can lead to depletion of oxygen in the water and promote the overgrowth of corals by macro-algae, especially when populations of grazing fishes and sea urchins are also reduced (e.g. by overfishing). For example, in Kaneohe Bay,

Hawaii, sewage discharges caused overgrowth of corals by algae, which was reversed when the sewage was diverted further offshore (Szmant 2002). Nutrient-stimulated growth of phytoplankton can greatly reduce light penetration to light-dependent corals (Australian Institute of Marine Science 2004; UNEP-WCMC 2006).

Industrial Pollution: Edinger et al. (1998) studied the impact of sediment, nutrient and/or industrial pollution on coral reefs in Indonesia. It was found that reefs subject to such pollution stresses showed a 30–60% reduction in species diversity. Reefs in Jakarta Bay are now entirely dead, and this has been attributed to the combined effects of industrial waste pollution and sewage. The exposure of reefs to oil pollution has also been shown to have damaging effects on corals (Brown 1997).

1.4.4.3 Coral Bleaching and Ocean Acidification

Increases in sea temperature due to climate change have led to ever more frequent coral bleaching events over the past 3 decades or so (see Chapter 5, Section 5.2.2). Bleaching describes the paling in coral colour that occurs when the symbiotic algae which live within the coral are lost in response to an increase in sea temperature. Coral bleaching is often temporary but can be permanent and cause coral death.

Increasing ocean acidity as a result of increases in atmospheric carbon dioxide could result in losses of coral reefs in the future (see Chapter 5, Section 5.8).

1.4.4.4 Disease

Coral disease outbreaks and the number of new coral diseases recognized have increased dramatically since the 1990s. Disease has affected many coral species; for instance, over 150 species from the Caribbean and Indo-Pacific alone (Australian Institute of Marine Science 2004). In the Caribbean, two of the most dominant reef-building corals have largely disappeared as a result of outbreaks of white band and white pox diseases. The precise causes of the recent increase in coral diseases are unknown. One study suggested that an increase in macro-algae on reefs could be a contributory factor (Nugues et al. 2004). Results from another study suggested that an increase in nutrient concentrations can increase the severity of coral diseases because the disease organisms can utilise the nutrients and become more virulent as a consequence (Bruno et al. 2003). It has also been noted that the increased incidence of coral diseases may be due to corals being physically debilitated following repeated bleaching events (Szmant 2002).

1.4.5 Reducing Threats and Conserving Coral Reefs

Coral reefs are not well suited to large-scale extractive exploitation and are only capable of supporting low-level harvests (Roberts 1995). The global decline of coral reefs necessitates that there should be combined actions to reduce all human threats to reefs (Pandolfi et al. 2005). Such action is urgently needed. In addition, a further essential tool in coral reef conservation is the establishment of no-take marine reserves. No-take marine reserves are areas of the marine environment in which all forms of extraction and disturbance by humans is permanently prohibited (see Chapter 7). The approach of implementing no-take marine reserves is increasing in coral reef management. For example, in 2004, no-take areas on the Great Barrier Reef, Australia, were increased from less than 5% of the total area to 33%, with a concurrent focus on improving water quality. Bellwood et al. (2004) suggested that this provides a good model. However, the authors noted that the proposal of the USA to incorporate 20% of its reefs as no-take areas by 2010 is a case of "too little, too late". Hughes et al. (2003) suggested that at least 30% of the world's coral reefs should be no-take areas, to ensure long-term protection of exploited stocks.

A global review of Marine Protected Areas (MPAs) concluded that there were 980 MPAs which incorporated coral reefs, together covering 18.7% of the world's coral reef habitats (Mora et al. 2006). However, of this 18.7%, 5.3% lay inside extractive MPAs, 12% lay inside multipurpose MPAs while only 1.4% lay inside no-take MPAs/ marine reserves. Furthermore, it was apparent that the establishment of MPAs was rarely followed by good management and enforcement. For instance, less than 0.1% of the world's coral reefs were within no-take MPAs that had no poaching. Even if protected by a reserve status, many coral reefs are also subjected to threats from outside their boundaries such as overfishing, pollution, sedimentation and coastal development. An assessment of such risks to coral reefs showed that less than 0.01% of the world's corals lie within no-take MPAs/marine reserves with no poaching and a low risk of other threats (Mora et al. 2006). This underlines the need for both the establishment of fully-protected marine reserves and action on other anthropogenic threats.

Research has shown that the establishment of marine reserves on coral reefs is beneficial in terms of increased species richness (more species) and their abundance. For example, East African reefs that had been protected for several years had higher species richness and abundances of some commercially important species compared to fished areas (McClanahan and Arthur 2001). In addition, reefs that had been protected for over 25 years had more and rarer species than reefs protected for less than 10 years, or fished areas. Another study on parts of the Great Barrier Reef, Australia, which had designated 'reserve' status for 12–13 years showed that the abundance of coral trout (*Plectropomus* spp.), the major target of hook and line fisheries in the region, had increased significantly compared to pre-reserve levels (Williamson et al. 2004).

There is an expectation that reserves will ultimately result in increased migration of fish outside the reserve boundaries and may therefore benefit fisheries near to reserve sites (the 'spillover effect'). There is some evidence from marine reserves in the Philippines (Russ et al. 2003) and Kenya (Kaunda-Arara and Rose 2004) that spillover can occur. Also, at the Soufrière Marine Management Area in St. Lucia,

Caribbean, the patchwork of closed areas of reef has sharply increased catches outside those closed areas (see also Chapter 7, Section 7.3.1). The areas open to fishing support as many fishermen as the whole reef did before the marine reserves were established. Much of the success of this network can be attributed to the full involvement of different stakeholder groups from the planning stages onwards (Renard 2001). On a cautionary note, Russ et al. (2003) commented that many species of coral reef fishes are very long-lived and so spillover from coral reefs may take decades to develop fully as the fish begin to increase in abundance and size.

In Fiji, nine out of ten people live in coastal communities and are dependent on the surrounding sea for their livelihoods. However, they are facing problems of diminishing marine resources. Local communities are now tackling the problem by returning to their traditional practice of establishing *tabu* or closed areas and combining this with modern conservation techniques to help rebuild stocks (Gell and Roberts 2003). In November 2005, local chiefs of Fiji's Great Sea Reef established five Marine Protected Areas with permanent no-take 'tabu zones'. This is an important step toward meeting the nation's commitment to build a Marine Protected Area network protecting 30% of Fijian waters by 2020. The Great Sea Reef, locally known as Cakaulevu, stretches over 200 km in length and is one of the longest barrier reefs in the world (aside from the Great Barrier Reef). It is home to thousands of marine species as well as being an important fishing ground for local communities (MPA News 2005).

The establishment of marine reserves may lead to overall economic benefits. In the domain of tourism (and ecotourism), studies have predicted that increased fish numbers and sizes fostered by marine reserves in the Turks and Caicos Islands would increase the economic viability of the reserves (Rudd and Tupper 2002). In Autralia, tourism is the most economically valuable industry in the Great Barrier Reef Marine Park. The park is divided into different zones including a large number of no-take zones. Tourism expenditure in 1999 was AU$4269 million, far exceeding the gross values of recreational fishing (AU$240 million) and commercial fishing (AU$119 million) (Commonwealth of Australia 2003). A further example comes from Apo Island marine reserve in the Philippines. Here it is estimated that the initial US$75,000 investment in the reserves now yields an annual return somewhere between US$31,900 and US$113,000 taking into account increased fish yields outside the marine reserve, and other reserve-generated income such as increased local dive tourism (White et al. 2000).

The placement of networks of marine reserves involves primarily biological evaluation and secondly socio-economic assessment (Roberts et al. 2003). In identifying selection criteria for marine reserves, Roberts et al. (2003) suggested that, once biogeographic regions have been identified, all habitats within each region must be represented in a network of reserves. For example, within the biogeographic region of a coastal zone, marine reserves supporting and linking different habitats such as coral reef, mangrove, seagrass bed, estuary and saltmarsh should be established. Coral reefs were identified as vulnerable habitats in particular need of protection. The need for action in establishing marine reserves is urgent. Biologically, several suggestions have been made regarding the selection of appropriate areas of coral reefs. Roberts et al. (2002) identified a number of important coral reef marine diversity hotspots based on protecting areas of high endemism (species restricted in their range). It was noted that 14 of the 18 centres of endemism identified were also adjacent to terrestrial biodiversity hotspots and, therefore, that extending terrestrial conservation efforts seaward in those regions could be an effective conservation strategy. Hughes et al. (2002) noted the importance of protecting the connectivity of widely dispersed reef species as well as peripheral areas that have high numbers of endemic species. Beger et al. (2003) suggested that rare species must also be considered when selecting protected sites.

1.5 Mangroves

Mangrove trees grow along coastlines in the inter-tidal zone between land and sea where they form mangrove forests (Fig 1.3). Geographically, mangrove forests are located between latitudes of 25° North and 30° South of the equator (Valiela et al. 2001). Mangroves support numerous species and serve to protect coastlines from storms but, despite their importance, a substantial proportion of mangrove forests have been lost due to human activities in recent decades. Valiela et al. (2001) esti-mated that 35% of the original area of mangrove forests has been lost in the last 2 decades. Other estimates suggest that more than 50% of the world's mangroves have been removed and that, whereas mangroves once occupied about 75% of tropical coastlines, they now only occupy around 25% (Rönnbäck 1999). Table 1.1 shows loss of mangrove areas between 1980 and 2000 for five world regions.

It has been estimated that there are approximately 175,000 km^2 of mangrove forests remaining along the world's shorelines, although there is some uncertainty in the calculations (Valiela et al. 2001). The largest proportions of mangroves occur in Asia and the Americas. Countries with the greatest areas of mangroves are Indonesia (42,500 km^2), followed by Brazil (13,400 km^2), Nigeria (10,500 km^2) and Australia (10,000 km^2) (Valiela et al. 2001).

Table 1.1 Estimate of mangrove area in different world regions for the years 1980 and 2000 and loss in area between 1980 and 2000 (Adapted from FAO 2003)

Geographical area	Mangrove area estimate (ha) in 1980	Mangrove area estimate (ha) in 2000	Loss in mangrove area (ha) between 1980 and 2000 (percentage loss in brackets)
Africa	3,659,322	3,350,813	308,509 (**8.4%**)
Asia	7,856,500	5,832,737	2,023,763 (**25.7%**)
Oceania	1,850,067	1,526,934	323,133 (**17.5%**)
North and Central America	2,641,289	1,968,407	672,882 (**25.5%**)
South America	3,801,600	1,974,300	1,827,300 (**48.0%**)

Fig. 1.3 Mangroves showing aerial roots, La Tola, Ecuador (Greenpeace/Takeshi Mizukoshi)

1.5.1   Biodiversity of Mangroves

All mangroves are comprised of salt-tolerant species, although there are a number of different ecological types of mangrove due to local variations in topography and hydrology (Rönnbäck 1999). A total of 69 mangrove species have been documented worldwide with the widest variety of species occurring in South East Asia (Field 2000). With regard to the forests themselves, Field (2000) noted that "*a mangrove forest often possesses a strange and convoluted beauty and it flourishes in conditions of heat, salinity and oxygen-starved mud that would overwhelm other terrestrial plants*".

Mangrove forests support extensive populations of birds, fish, crustaceans, microbes and fungi, as well as reptiles and mammals (Field 2000). Many mangroves support a high species richness of fish. For example, 117 fish species were recorded in the Matang mangrove waters of Malaysia, 260 species in Vietnamese mangroves and 400 species in the Sundarban mangrove forest of Bangladesh. The abundance of fish, crustaceans and molluscs has also been reported to be high in mangroves (Islam and Haque 2004; Rönnbäck 1999). Some mangroves support endangered species; for instance, the Milky Stork (*Mycteria cineres*), the crab-eating frog (*Rana cancrivora*) and leaf monkey (*Presbytis cristata*) in South East Asia (Field 2000), the manatee (*Trichechus manatus latirostrus*) in Florida, threatened Bengal tigers (*Panthera tigris tigris*) in India and Bangladesh, and plants such as the rare *Bulbophyllumrare* and other orchids in Singapore (Valiela et al. 2001).

Mangroves provide a rich source of nutrients for many invertebrates and fish that inhabit them (Field 2000; Rönnbäck 1999), as well as exporting nutrients that support other near-shore food webs such as shrimps and prawns (Valiela et al. 2001).

Mangroves also function as predation refuges for the larvae and juveniles of some marine species (Field 2000; Rönnbäck 1999), and although not many fish are permanent residents in mangroves, many marine species use mangroves as nursery areas for their larvae and juvenile life-stages. This includes, for example, many commercially important shrimp and crab species and offshore fishes (Islam and Haque 2004). A recent study showed that mangroves in the Caribbean acted as important nursery areas for some coral reef fish and may increase survivorship of young fish (Mumby et al. 2004). Where coral reefs were connected with mangrove habitat, the abundance of several commercially important species was more than doubled compared to reefs that were not near to mangroves. The study also suggested that the largest herbivorous fish in the Atlantic, the rainbow parrot fish (*Scarus guacamaia*), is dependent on mangroves and may have suffered local extinction due to loss of mangrove habitat.

1.5.2   Protection of Neigbouring Environments

In addition to being important ecological habitats, mangroves also help to stabilise coastlines and reduce erosion. They protect against floods and can mitigate the impact of severe tropical storms (Field 2000; Rönnbäck 1999). For instance, in the province of Phang Nga in Thailand, the presence of mangrove forests and sea grass beds significantly mitigated the impact of the 2004 tsunami (UNEP 2006b). In Bangladesh, China and Vietnam, mangroves have been planted for the purpose of preventing storm damage (Rönnbäck 1999). Mangroves also act to stabilise water quality in the coastal zone by trapping sediments, organic material and nutrients. The stabilising of water quality by mangroves is necessary for the functioning of nearby coral reefs (Rönnbäck 1999). Loss of mangroves can cause saltwater intrusion and deterioration of groundwater quality (UNEP-WCMC 2006).

1.5.3   Human Use of Mangroves

At a subsistence level, humans harvest fish, crabs, and shellfish from mangroves (Valiela et al. 2001). Coastal subsistence communities in many developing countries are very dependent upon sustainable harvests of seafood from mangroves (Rönnbäck 1999). Many mangrove villages practise the cultivation of fish and crabs within the mangrove waters. Such practice is usually small-scale and it is thought unlikely that permanent damage will be done (Field 2000). When developed on a commercial basis, however, the effects can be devastating (see Chapter 3, Section 3.2.1.2).

Commercial fisheries also rely on mangroves because mangroves support the development of many commercial species in their early life stages. For example, it has been estimated that 80% of all marine species of commercial or recreational value in Florida, USA, are dependent upon mangrove habitat for at least some stage in their lifecycle. For Fiji and India, it has been estimated that about 60% of commercially important coastal fish are directly associated with mangrove habitats. In eastern Australia, approximately 67% of the entire commercial catch is composed of mangrove-related species (Rönnbäck 1999).

The extent of mangrove forests has been shown to be linked to catches by fisheries, such that the greater the areas of mangrove cover, the higher the catches of fish. For example, a study on the south-east coast of India reported that mangrove-rich areas provided higher catches of shellfish and fish than mangrove-poor areas (Kathiresan and Rajendran 2002). Baran and Hambrey (1998) cited research on the Gulf of Mexico where it was shown that higher coastal vegetation, mainly mangrove, had a positive effect on commercial fish catches. Studies in Indonesia, the Philippines and Australia have also reported a positive effect of mangroves on shrimp catches (Baran and Hambrey 1998).

In addition to seafood, a variety of products are harvested on a subsistence level from mangroves. Wood is collected for construction, fuel and charcoal, while tannins are harvested for preservatives and dyes. Honey is gathered on a subsistence basis in numerous countries. Mangrove leaves are gathered in several countries to provide fodder for livestock. Medicinal plants are also collected from mangroves (UNEP-WCMC 2006). Where products are gathered for subsistence, villagers often appreciate the dangers of over-exploitation because the tradition of living with mangroves is strong. However, commercial exploitation by outside interests can cause massive disruption of the mangrove ecosystem (Field 2000) (see Section 1.5.4).

1.5.4 Threats to Mangroves

Large-scale destruction of mangroves is a relatively recent phenomenon. The pressures of increasing populations, food production and industrial and urban development threaten mangrove habitats (Field 2000). Conversion of mangroves for commercial aquaculture, agricultural, industrial and tourist facilities and industrial-scale forestry practices have led to dramatic losses of mangroves (Valiela et al. 2001). Field (2000) noted that mangroves have often been considered as wastelands by governments and planners whose approach has been to drain them and fill them in.

Valiela et al. (2001) suggested that the bulk of the increasing losses of mangroves on a global basis has been as a result of commercial aquaculture practices, in particular, shrimp farming (see also Chapter 3, Section 3.2.1.2). This practice results in destruction of mangroves to make shallow dyked ponds. There is typically a boom and bust cycle, as individual shrimp ponds rarely have a lifespan of more than 5–10 years. The practice is to then move on to destroy more mangroves, leaving abandoned unproductive land behind (Valiela et al. 2001; Rönnbäck 1999).

Mangroves, like other forests, play an important role in the global carbon cycle. The massive 35% losses of mangrove forests over the past 2 decades has resulted in the release of large quantities of stored carbon to the atmosphere (UNEP-WCMC 2006).

1.5.5 Reducing Threats and Conserving Mangroves

As noted above, mangroves are a vital habitat in the lifecycles of many marine species, and they provide important coastal protection. Subsistence harvests from mangroves are important to numerous local communities. Field (2000) suggested that the challenge for the 21st century is to ensure the well-being of subsistence dwellers who depend on mangroves, and to ensure that mangroves continue to grace the coastal shorelines of the world. Apart from sustainable subsistence harvesting from mangroves, another non-destructive use which can yield benefits for local people is ecotourism. This provides an incentive to preserve mangroves in their natural state and is viable as long as the numbers of people visiting does not itself cause habitat degradation (Field 2000).

The recent destruction of mangroves, largely by commercial enterprises, is clearly a threat to the ecological services they provide and to subsistence harvesting and commercial capture fisheries. A review of the current status of mangroves by the United Nations Environment Programme suggested that political will and concerted action are now needed for the protection of mangroves (UNEP-WCMC 2006). According to literature cited by UNEP (2006a), about 9% of mangroves lie within marine protected areas (MPAs). However, many more MPAs and marine reserves are essential requirements for mangrove conservation. Because of the connected functions between mangroves, reefs and seagrass beds, it has been suggested that efforts should be made to protect connected corridors of these ecosystems (Mumby et al. 2004).

1.6 Seagrasses

Seagrasses comprise a group of about 60 species of underwater marine flowering plants (Green and Short 2003). Seagrasses grow submerged in shallow marine and estuarine environments along most coastlines of all continents except for areas north of the Arctic Circle and south of the Antarctic Circle (Phillips and Durako 2000). Seagrasses often cover extensive areas, where they are referred to as seagrass beds or seagrass meadows. Although the number of seagrass species is fairly limited, seagrasses provide habitat for numerous other marine plant and animal species. It is the high biological productivity and physical complexity of the environment they provide which supports a highly diverse range of associated species.

Seagrass beds serve as feeding grounds and nursery habitat for many marine species including many commercially important species (Short and Neckles 1999). However, increasing coastal development over the past several decades has led to losses of seagrass throughout the world and is now a serious cause for concern (Phillips and Durako 2000).

1.6.1 Biodiversity of Seagrasses

Seagrasses themselves vary in structure, ranging from strap-like blades of eelgrass (*Zostera caulescens*) in the Sea of Japan at more than 4 m long, to the tiny 2–3 cm rounded leaves of sea vine (e.g. *Halophilia decipiens*) in Brazil's tropical waters (Green and Short 2003). Seagrasses may grow in stands of one species or may grow as mixed species. Seaweeds (macroalgae) may also live amongst the seagrasses and many species of algae have been found to grow on the seagrasses.

The plant-life of the seagrass beds provides both shelter and food for many animal species. The animal community of seagrass beds has been divided into several components: animals which live on the leaves or rest on them, animals which attach to the stems and roots of seagrasses, animals which are mobile and swim through the leaf canopy, such as fishes (Fig. 1.4), and animals which live in or on the sediment. Some animals spend their whole lifetime in the seagrass beds. Many others pass their juvenile life stages in the seagrass meadows, while others feed there during a portion of the day or during migratory movements. Thus the animal life of seagrass beds is made up of both permanent and temporary residents (Phillips and Durako 2000). The many species of organisms that are associated with seagrass beds include sponges, hydrozoans, sea anemones, solitary corals, worm-like

Fig. 1.4 *Coris julis* over a *Zostera* seagrass bed near Kas, Turkey (Greenpeace/Roger Grace)

animals, crustaceans, molluscs, echinoderms, tunicates, fishes, the green turtle and certain waterfowl and wading birds. Seagrass beds are also critically important food sources for two marine mammal species, the manatee (*Trichechus* spp.) and dugong (*Dugong dugon*) (Green and Short 2003).

As foraging grounds, some animal species feed on the seagrasses themselves while others feed on animals that inhabit the meadows (Jackson et al. 2001b). In addition, research has shown that senescent sloughed off seagrass leaves are decomposed by bacteria and that this detritus forms the basis of diverse food chains (Phillips and Durako 2000). Seagrass detritus may also enrich nearby unvegetated sediments, thereby releasing food to other fish in the vicinity. There is evidence to suggest that seagrass detritus may represent an important food input to coastal fisheries (Jackson et al. 2001b).

As nursery grounds, seagrass beds provide a place of both food and shelter. High concentrations of juveniles and larval stages of marine species have been identified in seagrass beds, including juveniles of fish and shellfish which are important to fisheries (Jackson et al. 2001b; Phillips and Durako 2000). Seagrasses are also thought to function as important nurseries for many coral reef fishes. For instance, a recent study showed that seagrass beds in some areas of the Caribbean provided important nursery habitat for a threatened species, the Indo-Pacific humphead wrasse (*Cheilinus undulates*) (Dorenbosch et al. 2006).

1.6.2 Protection of Neighbouring Environments

Beds of seagrasses serve to stabilise shorelines and reduce wave impacts (Beck et al. 2001). They stabilise sediments because of their interlacing rhizome/root mat. They have been reported to remain intact even through high wind and wave action during hurricanes in the Caribbean (Phillips and Durako 2000). In the province of Phang Nga in Thailand, the presence of seagrass beds was reported to have mitigated significantly the impact of the 2004 tsunami (UNEP 2006b).

1.6.3 Threats to Seagrasses

There is increasing concern regarding worldwide losses of seagrass beds. Losses have been due mainly to coastal development, during the past several decades. According to Phillips and Durako (2000), extrapolation of documented losses of seagrasses suggested that over 1.2 million hectares were likely to have been lost globally throughout the 1990s. Several reports have suggested that loss of seagrass habitat has been associated with declining fish catches (Jackson et al. 2001b; Phillips and Durako 2000).

Threats to seagrasses include dredging operations, reduced water clarity from nutrient and sediment inputs, and pollution. Dredging can be very damaging because not only is the entire plant removed, but also the sediment. Dredging can also cause turbid water that hinders plant growth. For example, continuous maintenance dredging at Laguna Madre, Texas, caused the loss 14,000 ha of seagrasses due to increased turbidity. Another dredging-type activity is the dragging of fishing nets and rakes across beds to collect shellfish. In some areas such activity has been found to cause significant damage to seagrasses. Boat propellers have also been found to cause damage, especially in shallow tropical waters. Pollution from sewage, agricultural chemicals and industrial chemicals can all be a threat to seagrasses. There may also be negative impacts from oil drilling and oil from shipping (Phillips and Durako 2000). In many cases, seagrass declines have been linked to multiple stresses acting together, but only in a few places are measures being implemented to address these threats (Green and Short 2003).

Rising sea temperatures due to climate change could directly threaten seagrass by altering growth rates and other physiological functions of these plants and by shifting their distribution. Sea level rise would also be likely to cause changes in the distribution of seagrasses (Short and Neckles 1999).

1.6.4 Reducing Threats and Conserving Seagrasses

Phillips and Durako (2000) suggested that activities leading to the decline of seagrasses should be stopped in order to preserve our national heritage and its resources. It was also suggested that legislation is needed which will guarantee the conservation and restoration of seagrasses on a global basis. The designation of areas encompassing seagrass beds as no-take marine reserves will be essential to the protection and conservation of these ecosystems. According to Green and Short (2003), no marine protected areas have yet been designated solely for the protection of seagrasses. However, the optimal design of marine reserves necessitates the inclusion of different habitats (Roberts et al. 2003) and, in the tropics for example, some marine reserves include seagrasses as well as other habitats such as reefs and mangroves. Indeed, because of the interconnected functions of coral reefs, mangroves and seagrass beds, it has been suggested that connected corridors of these habitats should be protected together (Mumby et al. 2004). Dorenbosch et al. (2006) suggested that measures aimed at conserving certain threatened species should be aimed at seagrass beds and mangroves that are likely to function as juvenile habitats.

Chapter 2
Fisheries

Abstract In the 1950s and 1960s, industrialisation of fisheries took off and there was a massive increase in global fishing effort and concurrently an increase in catches. Unfortunately, this led to widespread overfishing and eventually to the collapse of many fish stocks. Figures from the United Nations Food and Agricultural Organisation show that, despite further advances in fishing technologies, global catches have declined slowly since the late 1980s. Figures show that in 2005, 17% of stocks were classified as over-exploited and 7% as depleted. About half (52%) of fish stocks are classified as fully exploited, which means they are at, or close to, their maximum sustainable limits with no room for further expansion and even at the risk of decline if not properly managed. In the majority of cases it is overfishing that has led to stocks becoming over- or fully exploited. Recent studies have highlighted another concern stemming from over-exploitation by fisheries – that of a dramatic decline of many predatory fish across the world. In many cases there have been declines in predator stocks by about 90%.

Keywords Fisheries, overfishing, whaling, seabirds, turtles, by-catch, bottom trawling, beam trawling, longlining, ecosystem impacts.

A particularly unsustainable fishing method is the practice of bottom trawling. Bottom trawling is a method in which species living on, or close to, the seabed are fished, at depths of up to 2 km. It is highly destructive to marine habitats. Deep water fish can be especially vulnerable to over-exploitation through bottom trawling because they often have a slow growth rate, high longevity and are slow to reproduce. Older female fish tend to produce more eggs and, therefore, removing such fish can have a disproportionately large detrimental effect on the population.

Another fishing activity of concern is so-called industrial fishing, in which fish are taken for conversion to fishmeal or fish oil rather than for direct human consumption. Industrial fishing has led to the decline and collapse of some fish populations and impacts on seabirds, which consume these fish.

The incidental take of unwanted species by commercial fisheries, including seabirds, marine mammals and turtles is known as by-catch and is a serious problem.

M. Allsopp et al., *State of the World's Oceans*,
© Springer Science+Business Media B.V. 2009

These animals become entangled in nets or hooked accidentally by fishing gear and substantial numbers die as a consequence.

This chapter clearly shows that current fisheries management has failed to manage fish stocks in a sustainable way. To manage stocks sustainably, whole ecosystems need to be considered. The ecosystem approach maintains that extraction of all resources should be sustainable and should encompass the precautionary principle. A key tool in implementing the ecosystem approach is the establishment of a global network of fully protected marine reserves (see Chapter 7). Marine reserves are not only of benefit to aiding the protection and recovery of biodiversity but have also been shown to benefit fisheries. Outside of the marine reserves, an ecosystem approach requires the sustainable management of fisheries and other resources.

2.1 Introduction

Fishing became industrialised in the early 19th century with the advent of steam trawlers and later, after the First World War, with diesel engines. Further technological advances in fishing occurred after the Second World War, with the introduction of freezer trawlers, radar and acoustic fish finders. It was in the 1950s and 1960s when the industrialisation of fisheries really took off and there was a massive increase in global fishing effort and consequently an increase in catches. However, industrial-scale fishing was not sustainable. It inevitably led to over-fishing and subsequently to the collapse of fish stocks (Pauly et al. 2002).

2.2 State of the World's Fish Stocks

The Food and Agricultural Organization of the United Nations (FAO) have produced reviews on the state of the global fishery resources since the early 1960s (FAO 2005a). The latest figures show that in 2005, the total world fish production was 141.6 million tonnes. Of this, 84.2 million tonnes came from marine capture fisheries, 47.8 million tonnes came from aquaculture, and 9.6 million tonnes came from inland water capture fisheries (FAO 2007a).

Figures for 2004 showed that 75% of world fish production was utilised for direct human consumption and the remaining 25% for other products, in particular fishmeal and fish oil. With regard to fish for human consumption, of the estimated 106 million tonnes provided by capture fisheries and aquaculture together, 43% was provided by aquaculture (FAO 2007a).

An overview of FAO records shows that, in some areas, catches reached a maximum in the 1970s or early 1980s and have since declined. For marine capture fisheries, there has been an increase in the number of fully exploited or over-exploited fish stocks. Despite this, official figures showed that global catches seemed to continue increasing throughout the 1990s. However, it was later discovered that China had over-reported its marine fishery catches. Correction for this error revealed that catches had in fact declined slowly since the late 1980s by about 0.7 million tonnes per year (Pauly et al. 2002). A later analysis of figures was carried out by Zeller and Pauly (2005) in which consideration was also given to the decline in fishery discards in recent years. This analysis showed that, throughout the 1990s, global total fisheries catches may have declined by an estimated 1.6–2.3 million tonnes per year.

The most recent FAO figures show that about three quarters of marine fish stocks are now fully or over-exploited or depleted (FAO 2007a). Consequently, it was proposed that the maximum wild capture fisheries potential from the world's oceans has probably been reached. The FAO suggested that this "calls for a more cautious and closely controlled development and management of world fisheries". It was further noted that the situation seems particularly critical for some migratory, straddling and other fishery resources that are exploited solely or partially in the high seas (FAO 2007a). The FAO proposed that plans are needed to bring about replenishment of depleted stocks and to prevent further decline of stocks that are already exploited at, or close to, their maximum potential (FAO 2004a, 2007a). With regard to the large proportion of marine stocks that are now fully or over-exploited, it was suggested that, in the majority of cases, overfishing has been the prime cause of this problem. In some cases, environmental conditions have also contributed (FAO 2004a).

2.2.1 Summary of FAO Figures for Marine Capture Fisheries

Figure 2.1 depicts the status of world fish stocks in 2005 as defined by their degree of exploitation.

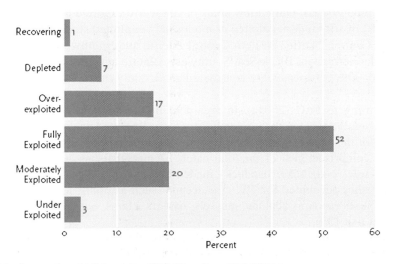

Fig. 2.1 Status of world fish stocks, 2005 (Data from FAO 2007a)

FAO figures on marine capture fisheries revealed that, between 1974 and 2003, there was a consistent downward trend in the proportion of fish stocks that had potential for expansion. Moreover, there was an increase in the proportion of over-exploited or depleted stocks during this time. Figures showed that in the mid 1970s, about 10% of marine stocks were classified as over-exploited or depleted but that this rose to close to a quarter (25%) of stocks in the early 2000s (FAO 2004a, 2005a). Figures from 2005 showed that of this estimated 25% of marine stocks, approximately 17% were classified as over-exploited, about 7% were classified as depleted and 1% were classified as recovering from depletion (FAO 2007a). The geographical areas with the highest proportions (46–60%) of over-exploited, depleted or recovering stocks were the Southeast Atlantic, Southeast Pacific, Northeast Atlantic and the high seas, particularly those in the Atlantic and Indian Oceans for tuna and tuna-like species (FAO 2007a).

About half (52%) of marine fish stocks in 2005 were classified as fully exploited. This means that half of catches were already at, or close to, their maximum sustainable limits, with no room for further expansion and even a risk of decline if not properly managed. The major fishing areas with the highest proportions (69–77%) of fully exploited stocks included the Western Central Atlantic, Eastern Central Atlantic, Northwest Atlantic, Western Indian Ocean and Northwest Pacific. Of the top ten species of fish that provide about 30% of the world's capture fisheries in terms of quantity, most species are considered to be either fully exploited or over-exploited. Therefore they cannot be expected to produce major increases in catches (FAO 2007a).

From 1974 to 2003, there was a downward trend in the proportion of under-exploited or moderately exploited marine stocks (FAO 2005a). By 2005, about a quarter of fish stocks were classified as moderately exploited (20%) or underexploited (3%) which means they may perhaps produce greater quantities of fish (FAO 2005a). Only a few areas of the world reported a high proportion (48–70%) of still under-exploited or moderately exploited stocks, including the Eastern Central Pacific, Western Central Pacific and Southwest Pacific, while in the Mediterranean, Black Sea, Southwest Atlantic and Eastern Indian Ocean about 20–30% of stocks were considered moderately or under-exploited (FAO 2005a).

According to FAO (2005a), in regard to the composition of marine-based catches, small pelagic fish (such as herrings, sardines and anchovies) accounted for 26% of the total catch in 2002. Larger pelagic fish (such as tuna, bonito and billfish) comprised 21% of the total catch. Demersal fishes (such as flounder, halibut, sole, cod, hake, haddock) made up 15% of catches. Miscellaneous coastal fishes accounted for 7%, crustaceans (such as crabs, lobsters, shrimps) 7%, molluscs (such as abalone, mussels, oysters, clams, octopus) 8%, and unidentified fish 13%.

Table 2.1 shows the top ten countries responsible for harvesting the highest quantities of fish from inland and marine capture fisheries in 2004.

Table 2.1 Top ten producer countries for inland and capture fisheries in 2004 (FAO 2007a)

Country	Production (million tonnes)
China	16.9
Peru	9.6
USA	5.0
Chile	4.9
Indonesia	4.8
Japan	4.4
India	3.6
Russian Federation	2.9
Thailand	2.8
Norway	2.5

2.3 Fishery Collapses and Declines of Marine Fish

In the 1950s and 1960s there was a huge increase in global fishing effort concurrent with improved technology and a consequent increase in catches (Pauly et al. 2002). In time, fisheries also extended their range from continental shelves to deep water habitats (Koslow et al. 2000). The increase in global fishing eventually led to declines and collapses in many fish stocks.

The collapse of a fishery can be defined as a sustained period of very low catch values occurring after a period of high catch values. Analysis of catch records between the years 1950 to 2000 indicated that collapses occurred in 366 fisheries out of a total of 1,519 (Mullon et al. 2005). This amounts to nearly one fishery in four collapsing over the 50-year period. During this time, the figures revealed that there was no sign of improvement in overall fisheries management to prevent collapses. Smaller fisheries and smaller stocks appeared to be more sensitive to collapse than larger ones. Demersal (bottom-dwelling) species were more susceptible than pelagic (open water) species. It has been predicted that there would be likely to be further collapses of many more fisheries in the future and that these would not always come with detectable warning signs (Mullon et al. 2005).

2.3.1 Collapse of Peruvian Anchovy and Atlantic Cod Fisheries

The first collapse of a fish stock which had global repercussions was the Peruvian anchovy (*Engraulis ringens*) in 1971–1972 (Pauly et al. 2002). This fishery was the largest in the world and it was brought to commercial extinction before later recovering (Goñi 2000). The cause was a combination of overfishing and the environmental impacts of an El Niño event (Goñi 2000; Pauly et al. 2002). Pauly et al. commented that, at the time, the collapse was blamed on the environmental effects rather than overfishing. This permitted 'business as usual' to go on in other fisheries. Subsequently, this action led to the beginning of an overall decline in total catches from the North Atlantic.

Over-exploitation in the Northwest Atlantic off the east coast of Canada resulted in perhaps the best known example of a fisheries collapse – that of the Atlantic cod (*Gadhus morhua*) fishery off Newfoundland. This area of the Northwest Atlantic had been fished for 500 years (Lear 1998). The decline in stock began in the 1960s (Haedrich and Barnes 1997) but there was respite in 1977 when Canada declared a 200-mile EEZ (Exclusive Economic Zone) which excluded most foreign fishing. However, in the late 1980s there was increased pressure from Canada's own fleet, and cod stocks collapsed in 1991 (Hilborn et al. 2003). In 1992 a moratorium was imposed and the great northern cod fishery was closed to commercial fishing (Schrank 2005). Figure 2.2 depicts the level of catches of the Grand Banks cod fishery from 1960 onwards. The closure in 1992 caused a major social upheaval, since 20,000 jobs were lost and the economy of Newfoundland was severely damaged (Hilborn et al. 2003). The cause of the collapse has been attributed primarily to overfishing in the late 1980s and early 1990s, with the consequence that both the numbers and size of cod decreased (Haedrich and Barnes 1997; Myers et al. 1996a, b). Although the fishery has remained closed, Schrank (2005) noted that there were no signs of recovery of the offshore cod in this area.

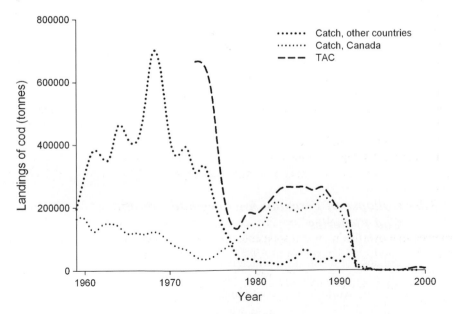

Fig. 2.2 Depletion of cod fishery on the Grand Banks (NAFO Divisions 2 K + 3 KL), Canada, from 1959 to 2000 (North Atlantic Fisheries Organisation)

2.3.2 Declines in Predatory Fish

Several recent studies have focused on the impact of fishing on predatory (high trophic level) fishes (Christensen et al. 2003; Devine et al. 2006; Myers and Worm 2003). One study analysed data on predatory fishes in the North Atlantic from 1900 to 1999 (Christensen et al. 2003). Thirty-one species of fish were assessed, including well known species such as cod (*Gadus morhua*), dogfish (*Squalus acanthias*), haddock (*Melanogrammus aeglefinus*), hake (*Merluccius merluccius*), herring (*Clupea harengus*), mackerel (*Scomber scombus*), salmon (*Salmo salar*) and whiting (*Merlangius merlangus*). The study revealed that, over the past 50 years, the amount (biomass) of predatory fish had severely declined by approximately two thirds. When figures for the past 100 years were considered, it was evident that predatory fish stocks had declined to just 11% of their former levels. The authors concluded that predatory fish species were therefore declining rapidly and stressed that what happens to the predatory fish species serves as an indictor of what we do to the ocean. Figure 2.3 shows declines in spawning stock biomass since 1960 recorded for four important commercial species in the North Sea, including the predatory species cod, haddock and whiting.

Myers and Worm (2003) analysed data for predatory fish stocks in four coastal continental shelf regions and nine oceanic deep-sea regions. The study analysed figures from 1952, when industrialised fishing began, up to the year 2000. Fish species from coastal shelf regions included codfishes (*Gadidae*), flat fishes (*Pleuronectidae*), skates and rays (*Rajiidae*), while fish from oceanic areas included tuna (*Thunini*), billfishes (*Istiophoridae*) and swordfish (*Xiphiidae*). Analysis of data indicated that industrialised fishing typically reduced the amount (biomass) of fish communities by 80% within 15 years of their exploitation. Moreover, results showed severe declines had occurred in the total quantities (biomass) of these predatory fishes to about 10% of their pre-industrial fishing levels. Therefore, 90% of these predatory fishes have been lost since 1952. The loss was widespread globally and occurred in both in coastal shelf regions and in the open ocean. Hence, this massive decline in predatory fish stocks was apparent in entire communities of fish across widely varying ecosystems. Consequently, the authors suggested that it is both appropriate and necessary to attempt restoration of fish stocks on a global scale. Figure 2.4 depicts graphically the results of trawl surveys in the open oceans used in the Myers and Worm (2003) study and shows consistent and rapid declines of catch rates over time.

There has been a sharp decline in the numbers of some shark species. For example, a study on coastal and oceanic shark populations of the Northwest Atlantic showed that overfishing was responsible for their rapid declines (Baum et al. 2003). Over a period of 15 years from 1986, scalloped hammerhead (*Sphyrna lewini*) declined by 89%, white sharks (*Carcharodon carcharias*) by 79%, tiger sharks (*Galeocerdo cuvieri*) by 65%, thresher sharks (*Alopias vulpinus*) and bigeye thresher (*A. superciliousus*) by 80%, blue sharks (*Prionace glauca*) by 60% and whitetip shark (*Carcharhinus longimanus*) by 70%. The extent of the declines suggested that several sharks in the Northwest Atlantic are at risk of large-scale extirpation.

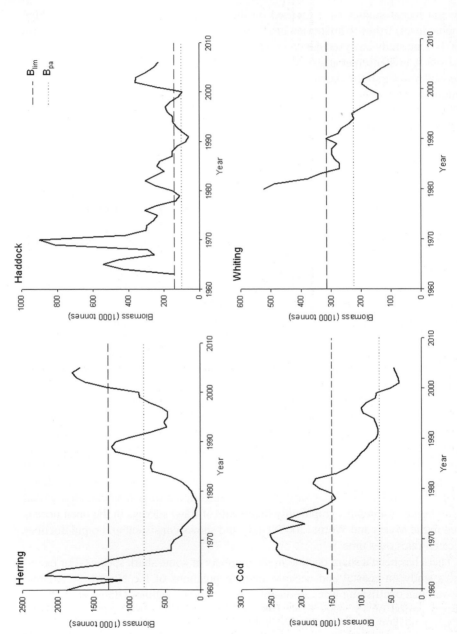

Fig. 2.3 Spawning stock biomass for four commercially important species in the North Sea (DEFRA 2005b), compared to ICES limit values (ICES 2007)

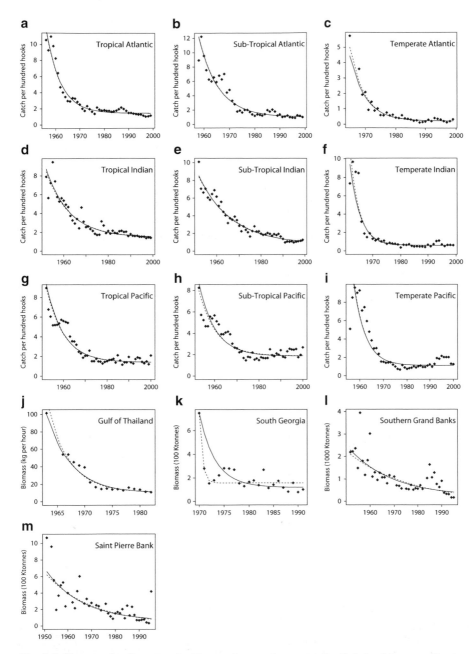

Fig. 2.4 Time trends of community biomass in oceanic ecosystems. Relative biomass estimates from the beginning of industrialised fishing (solid points) are shown with superimposed fitted curves from individual maximum-likelihood fits (solid lines) and empirical Bayes predictions from a mixed-model fit (dashed lines) (Myers and Worm 2003)

For another large predatory species, the Atlantic bluefin tuna (*Thunnus thynnus*), it has been concluded that the species has been undergoing heavy overfishing for a decade (Fromentin and Powers 2005). This species is managed by the International Commission for the Conservation of Atlantic Tunas as a western and eastern stock. It has been estimated that the western Atlantic bluefin tuna has declined by 80% or more since 1970 (Block et al. 2005). A global study of tuna and billfish from open ocean waters investigated the diversity (number of different species within a given area) of these fish over the past 50 years (Worm et al. 2005). Results showed that there was a decline in diversity of between 10% and 50% in all oceans. It was suggested that, while climate had year to year effects on populations, the primary cause for the long-term declines in variety for the tuna and billfish was overfishing. Figure 2.5 shows frozen tuna at a wholesale fish market in Japan, the world's largest consumer of tuna.

A study of Atlantic cod (*Gadus morhua*) stocks along the west coast of Sweden (in the eastern Kattegat and Skagerrak, North Sea) between 1978 and 1999 showed that stocks had become severely depleted due to unsustainable fishing pressure and were possibly on the verge of extinction (Cardinale and Svedäng 2004). There are also other species that have been fished to near extinction. These include the common skate (*Raia batis*), a species that was brought to the brink of extinction by fishing in the Irish Sea (Brander 1981). Skate are susceptible to the effects of fishing because they are slow growing, mature late in life and produce few offspring. The barndoor skate (*Raja laevis*) was once common in the northwest Atlantic but now appears to be near to extinction due to fishing pressure (Casey and Myers 1998). It was suggested that the only chance for survival of this species in the region would

Fig. 2.5 Frozen tuna being inspected prior to auction at the Tsukiji wholesale fish market, the largest fish market in the world, in Japan in 2006 (Greenpeace/Jeremy Sutton-Hibbert)

be to designate areas protected from trawling that were sufficiently large to allow for a self-sustaining population.

2.3.3   Declines in Other Species

Devine et al. (2006) investigated the decline in five species of deep-sea fish, some of them predatory, which inhabit the sea bottom of the continental slope in the North Atlantic Ocean. The species were the roundnose grenadier (*Coryphaenoides rupestris*), onion-eye grenadier (*Macrourus berglax*), blue hake (*Antimora rostrata*), spiny eel (*Notacanthus chemnitzi*) and spinytail skate (*Bathyraja spinicauda*). These species are vulnerable to the effects of overfishing because of their late maturation (in their teens), extreme longevity (up to 60 years of age), slow growth and low fecundity (produce few offspring). Between 1978 and 1994 there were large declines in the abundance of these five species, ranging from 87% to 98%. Apart from the spiny eel, all the species also decreased in mean size by 25–57%. The rapid decline in numbers of these five species qualify them for World Conservation Union (IUCN) categorisation as critically endangered.

In California, a marine invertebrate, the white abalone (*Haliotis sorenseni*), was exploited commercially from 1968. By 1996 the fishery was closed, after severe decline of the abalone to a level of 0.1% of its estimated pre-exploitation population size. The reduction in stocks occurred due to intense fishing. The white abalone has since been proposed for the endangered species list in the US. It is likely to become extinct in California within 10 years unless there is intervention to save it (Hobday et al. 2001).

In many countries, inadequate management of sea cucumber fisheries has resulted in severe overfishing, and depletion of stocks has become widespread. An example is the total collapse of the Sandfish sea cucumber (*Holothuria scabra*) at Abu Rhamada Island in the Red Sea due to overfishing from 2001 to 2003 (Hasan 2005).

2.3.4   Recovery Time for Fish Stocks Following Collapse

It has been noted that large species of fish are more likely to become depleted by fishing than smaller species. This is because small species typically have higher growth rates, mature earlier and have higher intrinsic rates of population increase (Badalamenti et al. 2002). One study has shown that, although fishery collapses may be reversible, the time needed for recovery appears to be considerably longer than previously thought (Hutchings 2000). Ninety fish stocks (consisting of 38 species) that had suffered prolonged declines over 15 years were studied. An assessment of their recovery over 15 years after their reduction showed that many gadids (for example, cod, haddock) and other non-clupeids (for example flatfishes) showed little, if any, recovery over the 15-year period. Greater recovery was only evident in clupeids (for example herring and sprat). This was most likely due to the fact that such species mature early in life and that fishing techniques for these species are selective. This is in contrast to bottom trawling for demersal fish, such as cod, haddock and flatfish, which does not permit high selectivity and may, therefore, hinder their recovery (Hutchings 2000).

2.3.5 Fishing Down the Marine Food Web

The trophic level of an animal describes the position it holds within the food web. For marine fish, the trophic level expresses the number of steps in the food web by which they are removed from the algae, which are situated at the bottom of the food web and assigned a trophic level of 1 (Pauly et al. 2002). Small zooplankton, which are mainly small crustaceans, consume the algae and are given a trophic level of 2. In turn, small pelagic fish such as sardines, herrings and anchovies usually consume a mixture of algae and zooplankton and are given a trophic level of 3. Larger fish that predate these small pelagic fish are assigned a higher trophic level, usually of 3.5–4.5. The trophic level of such fish is often not a whole number because they consume prey in several trophic levels. These larger fish are the typical table fish such as cod, tuna or halibut (Pauly and Watson 2003). They are usually larger in size than the fish they prey upon and they need more time to reach maturity and to reproduce, which makes them very susceptible to overfishing (Pauly et al. 2005).

Myers and Worm (2003) noted that, according to FAO statistics, there has been a consistent decline in the mean trophic level of fishery catches. This is of great concern because it implies the gradual removal of large, long-lived predatory fishes at the top of food webs and a switch to fishing for smaller, shorter-lived fish which are lower down the food web (Pauly et al. 2002; Pauly and Watson 2005). This process of sequential fishing to lower trophic levels has been termed 'fishing down the marine food web' (Pauly and Watson 2003; Pauly and Palomares 2005). The phenomenon occurs because fisheries are rarely closed following depletion of a target species but instead switch to catching other species (Goñi 2000). Pauly and Watson (2003) and Pauly and Palomares (2005) have suggested that the health and sustainability of fisheries can be assessed by monitoring trends in the average trophic levels. Once the trophic level of a fishery begins to fall, it indicates the dependence on ever-smaller fish and the collapsing of larger predatory fish stocks. It is already clearly evident that large predator fish in the world's oceans have been severely depleted (see Section 2.3.2). Large fish are generally more desirable than smaller fish because they yield higher economic returns (Goñi 2000).

Using FAO figures and other data sets, Pauly and Watson (2003) documented evidence that fishing down the food web is occurring in some fishing grounds, and it is happening on a global scale (Pauly and Palomares 2005). Regions where the trophic level had declined by 1 or greater between 1950 and 2000 included the North Atlantic, off the Patagonian coast of South America and nearby Antarctica, in the Arabian Sea and around parts of Africa and Australia. For instance, off the coast of Newfoundland, the average trophic level was 3.65 in 1957 and had dropped to 2.6 by 2000. During that time, the average sizes of fish caught dropped by a metre (Pauly and Watson 2003).

2.3.6 Impacts on Ecosystems of Overfishing and Fishing Down the Food Web

The direct impact on ecosystems of overfishing is the loss in abundance of the target species (Pauly et al. 2002). Near extinctions of two skate species as a result of overfishing have already been recorded (see Section 2.3.2) and it is possible that others may go unnoticed (Myers and Worm 2003). The ecological impacts of overfishing the world's predatory fish, to the extent that an estimated 90% have been lost, are bound to be widespread and possibly difficult to reverse (Myers and Worm 2003). In addition, fishing may selectively remove the larger, faster-growing individual fish from a population and it is possible that this may alter the genetic diversity of the population and hence its survival capabilities. It has been suggested that this phenomenon has implications for considering the ecological impact of fisheries in their management (Pauly et al. 2002).

An impact on marine ecosystems from the practice of fishing down the food web is that it will reduce the number and length of pathways that link food fishes to algae at the base of the food web. The result is that the food web may become simplified. This is important because a normally diversified food web allows predators to switch prey if their food source fluctuates with climatic effects, and hence compensate for environmental fluctuations.

It may be argued by some that fishing down the marine food web is a good thing because it releases other fish from predation and allows their numbers and hence catches to increase. However, such effects are seldom found in marine ecosystems. In actuality, predator fish help to maintain the complex food webs in which their role tends to support production of their prey. For instance, a predator may have a direct negative impact on its prey but also an indirect positive effect by consuming other predators and competitors of prey (Pauly et al. 2002).

2.4 Whaling and Declines of Cetaceans (Whales, Dolphins and Porpoises)

2.4.1 Historical Over-exploitation of Whales

The history of the commercial whaling industry is one of repeated over-exploitation, which took many populations of whales to the brink of extinction (Johnston et al. 1998). In the 17th century, Dutch and British fleets targeted the bowhead whale, clearing the Arctic Ocean east of Greenland and moving towards American waters. The whaling industry centred around Cape Cod targeted right whales and, subsequently, sperm whales. The whaling ships moved increasingly further south in the Atlantic and, rounding Cape Horn, began to exploit the stocks of sperm whales present in the Pacific Ocean.

The development of mechanically powered vessels in the early 20th century, together with harpoon guns and inflation lances to inflate carcasses with compressed air and keep them afloat, allowed the targeting of fast whales such as the blue, fin and sei whales. In addition, it allowed the scale of operations to be greatly expanded, ensuring that whale products were kept at a price low enough to sustain a market. After stocks became depleted in the northern seas, exploitation began in the Antarctic on the whale feeding grounds where the animals congregated. Norway, the pioneer country, was joined by the UK and, in the 1930s, by Japan and Germany. The initial targets of the fast catcher boats were the blue whales. By 1925, 2,500 were being taken annually. In 1926, the first factory ship, serviced by a fleet of fast catcher vessels was introduced. This innovation removed even the imperfect controls that could be exerted on a land-based industry. By 1930, the catch of blue whales had reached 30,000. Overall, between 1910 and 1966 over 330,000 blue whales were killed.

After the blue whales became scarce, fin whales were targeted and caught at a rate of almost 30,000 per year in the 1950s. Technological advances and a continuing market demand for whale oil ensured that whaling continued to be a growth industry (Knauss 1997). Much of the growth was in the Antarctic. About 2,000 whales were taken in the region in the 1907–1908 season, rising to 14,000 in 1927–1918. In 1937–1938, immediately before the Second World War, 46,000 animals were killed. This accounted for around 85% of the total number of whales caught in that year. Sei whales were increasingly targeted in the 1960s and minke whales in the 1970s.

The history of commercial whaling is represented in Fig. 2.6, showing the successive targeting and serial depletion of some of the Great Whales (Evans 1987). In total, over 1.5 million whales were killed in the 50 years after 1925.

2.4.2 The International Whaling Commission

The International Whaling Commission (IWC) was set up in 1946 to regulate commercial whaling and to conserve whale populations for the benefit of both present and future human generations. Its failure to manage effectively is well documented and led to the collapse of most global whale populations (Gales et al. 2005; Holt 2006). In 1972, the UN General Assembly called upon the IWC to implement a precautionary 10-year moratorium on all commercial whaling. The call was initially ignored by the IWC but was finally agreed upon in 1982 and the moratorium became effective in 1986. The time period of the moratorium was of undefined length, with a review set for 1990. The moratorium applied to all whale species

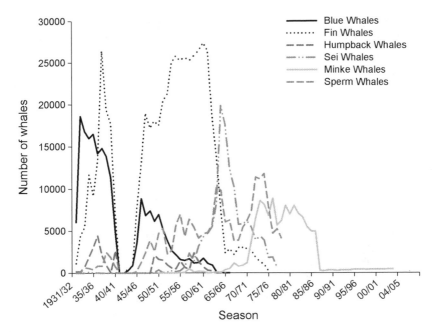

Fig. 2.6 Catches of Great Whales in thousands with season. The curves clearly show the sequential targeting of the most commercially valuable species, as populations of individual species were successively over-exploited (After Evans 1987)

for which the IWC accepts responsibility, that is, baleen whales (whales that filter food from water by plates of baleen) and, among toothed whales, the sperm whale, the northern bottlenose and the orca (Holt 2006). Many of the smaller cetaceans (small whales, dolphins and porpoises) are not covered by the moratorium. There are also two exceptions to the moratorium, namely aboriginal/subsistence whaling and so-called 'scientific whaling'.

In 1979, the IWC established a whale sanctuary in the Indian Ocean and closed the waters to commercial hunting. Another whale sanctuary was created in 1994 in the Southern Ocean. The aim of the sanctuaries was to provide protection for whales to aid in their long-term conservation (Morgera 2004).

2.4.3 Current Status of Commercial Whaling and Scientific Whaling: Threats to Cetaceans from Whaling

2.4.3.1 Norway

Norway initially stopped but resumed commercial whaling in 1993 and continues to catch large numbers of minke whales in the northeast Atlantic. In 2006, the quota was set at 1,052 minke whales. The Norwegian government claims that the hunt is small scale and traditional, yet Norway did not begin minke whaling until 1930. In reality, there is now little market for whale meat in Norway.

2.4.3.2 Iceland

After a hiatus of several years, Iceland initiated a hunt for whales in 2003 under the guise of 'scientific whaling'. Taking whales for scientific research purposes is allowed under Article VIII of the Convention. However, the whale meat from these hunts is sold commercially. The Scientific Committee of the IWC reviewed Iceland's scientific programme and subsequently the IWC asked Iceland not to conduct the programme. Indeed, whale experts from around the world have demonstrated viable alternatives to lethal research, which makes killing whales for science unnecessary (Clapham et al. 2007). In 2006, Iceland also announced a small commercial whaling programme authorising the killing of 30 minke and 9 fin whales. However, in 2007 the commercial hunt was stopped because of a lack of demand for the whale meat from the previous hunt.

2.4.3.3 Japan

Following the onset of the moratorium on commercial whaling set by the IWC, Japan continued whaling for 2 years but then stopped (Holt 2002). Japan began scientific whaling in 1987 in the Antarctic and subsequently expanded its programme into the western North Pacific in 1994. Gales et al. (2005) reported that, since 1987, Japan has caught 7,900 minke whales, 243 Bryde's whales, 140 sei whales and 38 sperm whales for scientific purposes. Figure 2.7 shows the transfer of two captured whales to a Japanese factory ship in the Southern Ocean in 2005. Plate 2.1 shows a whale finally captured after repeated harpooning. Presently, Japan has set the annual catch for scientific whaling at a maximum of 1,415 whales from seven different species (Kasuya 2007). And yet, by contrast, only 840 whales were killed globally by Japan for scientific research between 1954 and the moratorium. This discrepancy calls for an explanation of Japan's scientific rationale. Indeed, many members of the Scientific Committee of the IWC have consistently complained that such catches do not have sufficient scientific basis. Gales et al. (2005) noted that the strongest scientific argument in favour of lethal sampling is the collection of genetic samples for determining population structure. However, this can be carried out using non-lethal biopsy techniques. Very few peer-reviewed scientific papers have resulted from Japan's scientific whaling programme. Greenpeace has long expressed the view that Japan's scientific whaling programme is nothing other than an economic activity designed to supply whale meat for its country.

There is widespread concern because some of Japan's scientific whaling is conducted in waters of the Southern Ocean Whale Sanctuary. Consequently, there have been repeated calls by the IWC to stop whaling in this region, as yet to no avail

Fig. 2.7 Greenpeace witnesses the killing of whales in the Southern Ocean in 2005 by the Yushin Maru and the Kyo Maru No. 1 ships of the Japanese whaling fleet, and the transfer of the whales to the Nisshin Maru factory ship (Greenpeace/Jeremy Sutton-Hibbert)

(Gales et al. 2005). According to Gales et al. (2005), Japan's planned catches of humpback and fin whales in the Southern Ocean are of particular concern because the conservation status of these animals is, respectively, 'vulnerable' and 'endangered'. In addition, an apparently recent decline of Antarctic minke whales is of great concern to the Scientific Committee of the IWC (Kasuya 2007). Greenpeace has repeatedly taken action to highlight the activities of the Japanese whaling fleet in the Southern Ocean (Plate 2.2).

Plate 2.1 The Yushin Maru catcher ship of the Japanese Southern Ocean whaling fleet injures a whale with it's first harpoon attempt, and takes a further three harpoon shots before finally killing the badly injured fleeing whale (Greenpeace/Kate Davison)

Plate 2.2 Greenpeace crew-members use a modified fire pump in a small inflatable to obscure the view of the harpooner on the Yushin Maru No. 2 of the Japanese whaling fleet in the Southern Ocean, 2005 (Greenpeace/Kate Davison)

There are further threats to some small cetaceans from commercial hunting by Japan since these are not covered by the moratorium on commercial whaling. The total annual catch allowed for the small cetacean fishery is over 20,000 individuals. This quota was set in 1993, based on abundance estimates at the time and assuming an arbitrarily selected population growth rate of 2–4%. Sustainability has not been demonstrated and population numbers in the following years have not been studied. This is a conservation concern because some of the small cetacean stocks are very heavily hunted, or killed incidentally in fisheries. For example, concerns have been raised by the Scientific Committee of the IWC regarding populations of the short-finned pilot whale and striped dolphin, which are hunted by Japan. Concerns have also been raised about the plight of the Dall's porpoise. For the last 25 years, Japan's Dall's porpoise hunt has been the largest cetacean hunt in the world, with as many as 17,700 animals slaughtered each year. Continued hunting has been justified on the basis of abundance estimates from Japan's Fishery Agency which are more than 15 years old, and has consistently exceeded the IWC's recommended safe level of mortality. Consequently, populations have declined and Japan has resorted to killing female porpoises with nursing calves because they are easier targets. The calves are left to die but are not included in catch statistics (EIA 2006).

Presently, Japan refuses to cooperate with the Scientific Committee on the management of small cetaceans, a matter of extreme concern (Kasuya 2007). Some IWC commissioners now want to bring the hunting of small cetaceans under the control of the IWC, but others, such as Japan, reject such proposals.

2.4.4 Threats to Cetaceans from Fisheries

Today, cetaceans are facing a myriad of threats in addition to hunting, including climate change, toxic pollution, noise pollution, ship strikes and death or injury from entanglement/ingestion of marine debris (see Chapter 4, Section 4.5). Fisheries pose a major threat to cetaceans because many are caught and killed accidentally in fishery activities (see Chapter 2, Section 2.5.2).

A further threat to cetaceans is the reduction of their prey due to overfishing. As discussed in this chapter, commercial fishing has altered and degraded marine ecosystems such that many species are now fully exploited or over-exploited and much reduced in abundance. It is likely that large whales in certain areas have been affected by over-harvesting of fish and other marine species. For example, it has been suggested that the virtual removal of the North American Georges Bank stock of herring (*Clupea harengus*) by over-fishing in the 1960s removed a major prey source for local populations of baleen whales (Clapham et al. 1999). As Clapham et al. (1999) note, it has been suggested that this exploitation substantially changed the ecosystem dynamics and the feeding habits of baleen whales in the region.

An example of where overfishing is likely to have been a major contributing factor in the decline of a cetacean species is that of the population of short-beaked common dolphins (*Delphinus delphis*) inhabiting the Mediterranean. There is evidence that the species has undergone a dramatic decline in the central and eastern Mediterranean in recent years and the IUCN have suggested that conservation action is urgently needed to prevent extirpation of the species in the region (Reeves et al. 2003). A review of the status of the dolphins in the Mediterranean in 2003 concluded that prey depletion due to overfishing is one of the most plausible hypotheses explaining their population decline, although a range of other threats was also apparent (Bearzi et al. 2003). The review noted that a lack of sufficient food to maximize reproductive potential has been cited as possibly the most important regulator of population size in animals. In the Mediterranean, available data indicate that unsustainable harvesting has led to the decline of many fish stocks and there is evidence suggesting that marine food webs are being 'fished-down', i.e. exploitation focusing further and further down the food chain as predator species are progressively fished out. For example, fishing for small pelagic fish in the Alboràn Sea has increased dramatically in recent years, to the extent that this is suggested to cast doubt on whether the common dolphin will be able to persist even at current population levels (Bearzi et al. 2003).

Despite the likelihood that cetaceans may be suffering from depleted prey due to overfishing, the Japanese authorities have proposed that whales are consuming five times more fish resources than humans and that whales are therefore competing with humans for limited fisheries resources. The proposal was formulated by the Japanese Institute of Cetacean Research (ICR), the body that oversees the Japanese government's scientific whaling programme. The argument has, however, been countered by other scientists as highly flawed and naïve. For example, to a very great extent, the target fish of commercial fisheries do not coincide with those taken by whales. When this is taken into account, together with scientific figures for the assumed intake of food by whales, a re-calculation of the amount of fish consumed by whales turns out to be significantly lower than the ICR estimate. Furthermore, the greatest predation on fish populations in the majority of ecosystems studied is, in fact, largely from other predatory fish and not from whales (Johnston and Santillo 2000). In another study, Kaschner and Pauly (2004) showed that there is actually little overlap between where the majority of marine mammals and fishing fleets spend their time. For instance, at least 80% of the world's fish catch comes from regions where there is very little overlap with cetaceans. Furthermore, only about 1% of food taken by all marine mammal groups is consumed in areas with significant spatial or dietary overlap with fishery catches.

In response to the ICR claim, Holt (2006) noted that the 'whales-are-eating-our-fish' argument is a pretext for 'culling', to prevent the recovery of depleted populations and reduce others, a policy of deliberate, unsustainable whaling. Holt (2006) goes on to say that ICR documents contain stupendous mistakes of method and errors of calculation and grossly misleading 'conclusions'. As such, in regard to the ICR, Holt (2006) notes that "*An institute that produces such materials, and uses them to prop up a national strategy to further deplete, for more short-term profit, already stressed marine resources should not be treated by the international scientific community as a legitimate research body*". It is a fact that the central problem of depleted fisheries worldwide is one of consistently poor regulation and management, which has led to over-exploitation of stocks (Johnston and Santillo 2000).

2.4.5 Protecting Cetaceans

As discussed above, many cetaceans were heavily exploited in the past, leading to severe depletion of populations. Presently, while some species are still directly threatened by hunting, all cetaceans are now also exposed to a combination of threats from human activity which are likely to be impacting upon them synergistically – including by-catch in fisheries, prey depletion, toxic pollutants, and climate change. In this regard, the plight of cetaceans can be seen as symptomatic of what is happening to the oceans at large.

In order to protect cetaceans and aid the recovery of their populations, it is necessary to protect the marine ecosystems of which they are a part. One way to secure protection for the whole spectrum of marine wildlife from the smallest planktonic organisms to the largest whales is by establishing a significant proportion of the oceans as marine reserves (see Chapter 7). In addition, the creation of *specific* whale sanctuaries affords protection to cetaceans. One example is that, in 2002, Mexico declared a sanctuary of approximately 3 million square kilometres, giving protection to 21 cetacean species in the region. Both whale sanctuaries and marine reserves can also help to offer economic benefits in terms of the development of whale-watching programmes. When managed on sound ecological principles, whale watching can be sustainable and can bring significant economic rewards to coastal communities. It is now a thriving industry, with at least 87 countries running whale-watching operations (Hoyt 2001).

2.5 Fishing Methods of Concern

Overfishing using any fishing method is of ecological concern, as discussed in Section 2.3 of this chapter, but there are some methods of fishing whose impacts on marine ecosystems are of particular concern. These include bottom-trawling and industrial fishing. Bottom-trawling is a fishing method which involves towing trawl nets along the seafloor (as opposed to pelagic trawling where a net is towed higher in the water column), and is very destructive to habitats on the seafloor. Industrial fishing involves taking small pelagic fish for non-food purposes, and is often unsustainable. The consequences of bottom-trawling and industrial fishing are discussed in Sections 2.5.1–2.5.3 below. Other fishing methods of concern, including long-lining and seining, are discussed in relation to their impacts on by-catch of seabirds, turtles and marine mammals (Section 2.6).

2.5.1 Bottom-Trawling in the Deep Sea

If we could view the planet beneath the seas we would see that the landmass of the continents extends outwards in a 'continental shelf', typically a couple of hundred metres below the surface. The shelf then drops steeply several thousand metres to the vast abyssal plain, typically around 5 km deep. Superimposed on this overall topography are seamounts (underwater mountains that don't break the surface) together with small hills and undulations (McGarvin 2005).

As traditional fisheries on the continental shelf have declined, fisheries have moved increasingly to exploit deepwater habitats in which fish congregate, such as seamounts. Deepwater fisheries may be defined as fishing at depths of greater than 500 m (Koslow et al. 2000). Depths of up to 2 km can now be reached (McGarvin 2005). When species on, or close to, the seabed are fished, this is known as bottom trawling. Bottom-trawling equipment such as rockhoppers are used, which have rubber discs or steel bobbins that ride over obstructions such as boulders and coral heads that might otherwise snag the net. The recent use of bottom trawling equipment together with global positioning systems and fish finding equipment has allowed trawlers to operate in previously unfished waters. Bottom fishing can now be carried out on every sort of bottom type from subpolar to tropical waters (Watling and Norse 1998).

Bottom trawling is, however, highly destructive to marine habitats. Indeed, from the fishing industry to academia, there is consensus that bottom trawling is the most destructive of all fishing methods (Deep Sea Conservation Coalition 2005). The use of heavy gear which drags across the seabed causes massive collateral damage. Corals that afford protection and habitat for many creatures are ripped up and destroyed (see, for example, Fig. 2.8). Biodiversity is reduced. Numerous species are caught other than those being targeted, which are frequently dumped dead or dying back into the sea (McGarvin 2005). Bottom trawling is so destructive that it has been likened to having effects on the seabed that resemble forest clear-cutting

Fig. 2.8 Sample of fragile deep sea coral retrieved from the nets of a Lithuanian bottom trawler the ANUVA in the North Atlantic, 2004 (Greenpeace/Steve Morgan)

(Watling and Norse 1998). The International Council for the Exploration of the Seas (ICES) concluded that, while all deep-water fishing gear had some impact on the seabed, bottom-trawl fishing was the most damaging to deep-water corals and other vulnerable species (Deep Sea Conservation Coalition 2005).

2.5.1.1 Vulnerable Species

Bottom trawling kills seabed life forms by crushing them, by burying them under sediment and by exposing some bottom-dwelling fauna to predators. Life forms such as sponges, corals, bryozoans and polychaetes may be removed or damaged by bottom trawling (Watling and Norse 1998). This reduces habitat for a myriad of other species and food for others. For example, deep-sea corals, sponges and other habitat-forming species afford protection from currents and predators, act as nurseries for young fish and are feeding, breeding and spawning areas for numerous fish and shellfish species (Roberts and Hirshfield 2004). Plate 2.3 shows the stark contrast between a heavily trawled area of seabed off Northwest Australia and an untrawled area.

There is evidence of substantial damage to deep-water corals from bottom trawling off the coasts of Europe from Scandinavia to northern Spain, off both coasts of North America and on seamounts near Australia and New Zealand (Roberts and Hirshfield 2004). Off Norway, for example, it has been estimated that between a third to a half of deepwater *Lophelia pertusa* reefs have been damaged or destroyed by trawling. Similarly, widespread damage to deep-sea reefs off Ireland and Scotland has been reported. In regions off Norway and the UK, photographs show giant trawl scars up to 4 km long and it has been estimated that some of the reefs impacted are about 4,500 years old. In the waters off Atlantic Florida, it has been estimated that 90–99% of *Oculina* reef habitat has been reduced to rubble. Extensive damage of coral reefs on seamounts off Tasmania has been reported (Roberts and Hirshfield 2004). On two heavily fished seamounts off New Zealand, coral cover was reduced to 2–3% of the photographed areas compared with less fished seamounts, which often had 100% coral coverage (Clark and O'Driscoll 2003).

Corals can be hundreds to thousands of years old. There has been little long-term research on recovery rates of deepwater corals and sponge communities. However, since they have high longevity and slow growth it is likely that their recovery will be exceedingly slow (Roberts and Hirshfield 2004). In summarising the destruction of corals by bottomtrawling, the General Secretary of ICES made the following observation: "*Towing a heavy trawl net through a cold-water coral reef is a bit like driving a bulldozer through a nature reserve. The only practical way of protecting these reefs is therefore to find out where they are and then prevent boats from trawling over them*" (Roberts and Hirshfield 2004).

Deep water fish can also be especially vulnerable to over-exploitation through bottom trawling. This is because they often have a slow growth rate, high longevity and are slow to reproduce (Clark 2001). Maturity, and therefore breeding potential,

Plate 2.3 The seabed off Northwest Australia depicting (i) an untrawled area and (ii) a heavily trawled area (CSIRO)

may be delayed by over 20 years (Koslow et al. 2000). Older female fish tend to produce more eggs, and produce eggs with higher nutrition content that have much higher chances of survival in their early life stages. Therefore, removing older fish is often detrimental to the population. The low resilience of many deep-sea fish populations will result in decade- to century-long recovery times (Deep Sea Conservation Coalition 2005).

Koslow et al. (2000) suggested that a boom and bust cycle has characterised many individual deep-water fisheries, such that most stocks have become overfished or even depleted. An example of a deepwater fish exploited by bottom trawling is the orange roughy (*Hoplostethus atlanticus*), which has been fished intensively on seamounts and deep plateaux around New Zealand and Australia (Koslow et al. 2000). In New Zealand, initial catches from the early 1980s to 1990s were high but, in a familiar pattern, this was followed by reductions in catches and contractions in stock distribution (Clark 2001; Koslow et al. 2000) followed by seeking out new areas to fish. Serial depletion of populations around seamounts off New Zealand has occurred (Clark 2001). Stocks off seamounts around Tasmania were also fished down within several years. In addition, there has been progressive serial depletion of orange roughy stocks between Southeastern Australia and New Zealand. Newly discovered stocks were typically fished down to 15–30% of their original biomass within 5–10 years (Koslow et al. 2000). Koslow et al. (2000) suggested that it is doubtful whether the orange roughy fishery can be sustained at present levels, given the lack of further seamounts to be exploited.

Two other fish species, the smooth oreo (*Pseudocyttus maculates*) and the black oreo (*Allocyttus niger*) are fished in similar habitats to the orange roughy in Australia and New Zealand. There is considerable uncertainty as to whether current landings are sustainable (Koslow et al. 2000). In the North Sea, skates and rays, which have a high age at maturity, and are slow to reproduce, have disappeared from large areas due to intensive bottom-trawl fisheries in the North Sea (Wolff 2000).

The technique of bottom trawling of seamounts may threaten life forms which have not yet even been recorded by scientists. For example, a study of seamounts in Tasmania found about 300 species of fish and invertebrates, 24–43% of which were new to science, while 16–33% were restricted to the seamount environment.

2.5.1.2 By-catch

One of the ecological costs of fishing is the incidental catch of non-target species (Sánchez et al. 2004). By-catch from bottom trawling fisheries has been reported to be high. For example, a study on bottom trawl discards in the Northeastern Mediterranean from 1995–1998 reported that about 39–49% of the catch was discarded (Machias et al. 2001). Another study in the Northwestern Mediterranean between 1995 and 1996 found that bottom trawling catches comprised 115 species that were kept for the market and 309 species that were discarded. On average, discards accounted for one third of the catch by weight (Sánchez et al. 2004).

2.5.1.3 Countries Engaged in Bottom-Trawling

Those primarily responsible for bottom trawling are the fishers and fishing companies who moved into deep sea fisheries in search of new markets. In 2001, just 11 countries were responsible for 95% of the reported high-seas bottom trawl catch – Spain, Russia, Portugal, Norway, Estonia, Denmark, Faroe Islands, Japan, Lithuania, Iceland, New Zealand and Latvia (McGarvin 2005). Plate 2.4 shows a

Plate 2.4 Catch landed on board EU bottom trawler, the Ivan Nores, in the Hatton Bank area of the North Atlantic, 2005 (Greenpeace/K Davison)

typical catch of deepwater bottom-dwelling fish taken from the North Atlantic by a Spanish bottom-trawler in 2005.

2.5.1.4 Policy

Many of the deep-sea fisheries lie within the control of coastal states. However, much of the management of these deep-water stocks has been very poor and little attention has been given to the impact of heavy bottom-trawling gear on habitat. Furthermore, the search for new deep-sea stocks has extended onto the high seas, where there is either little or no management at all. In general, there has also been relatively little information published on the impact of bottom trawling in such habitats (McGarvin 2005). However, there is no doubt that it is highly destructive to marine habitats and that many life forms are vulnerable to this type of fishing.

To date, only a few areas have been afforded protection from bottom trawling. However, one way that the global marine environment could be protected from high seas bottom trawling is by a legally binding agreement under the control of the United Nations (UN). A number of countries, along with marine scientists and environmental groups, have been lobbying the UN to impose a moratorium on high seas bottom trawling. A UN General Assembly Resolution placing such a moratorium is the most immediate way to provide protection for vulnerable marine ecosystems. It would permit 'time out' to make proper scientific assessments of deep-sea ecosystems and to develop the policy solutions necessary to conserve these ecosystems well into the future (McGarvin 2005). The UN Secretary

General's advisory body on the implementation of the Millennium Project has recommended that "*global fisheries authorities must agree to eliminate bottom trawling on the high seas by 2006 to protect seamounts and other ecologically sensitive habitats and to eliminate bottom trawling globally by 2010*". Action is now needed to bring a moratorium into place.

2.5.2 Beam Trawling on the Western European Continental Shelf

Beam trawling is a method of bottom trawling in which the mouth of the net is held open by a solid metal or wooden beam. The beam is attached to metal plates at either end which slide over the sea bed as the net is towed along. When beam trawling is used to catch flatfish (rather then shrimps), tickler chains are added to the metal plates which disturb the upper few centimetres of sediment, and this increases the catch of the target species (Kaiser and Spencer 1996).

Beam trawling for flatfish began in the early 1960s and, for the next 2 decades, was largely restricted to the Netherlands and Belgium. By the late 1980s, beam trawling had expanded into the United Kingdom and Germany (Rijnsdorp and Leeuwen 1996). The beam trawl fishery on sole (*Solea solea*) and plaice (*Pleuronectes platessa*) is now the most important fishery in the North Sea. The practice of beam trawling in the North Sea more than doubled between 1971 and 1993 (Groenwold and Fonds 2000). In 2000, Groenwold and Fonds noted that almost 70% of the total nominal sole catches were landed by the Dutch beam trawlers. Figure 2.9 shows a UK registered beam trawler fishing in the North Sea.

Fig. 2.9 Beam trawler in North Sea (Greenpeace/Fred Dott)

It is of concern that the beam trawl fishery for flatfish produces large amounts of dying discards (Groenwold and Fonds 2000) and kills, damages or exposes large numbers of bottom-dwelling sea organisms that are in the path of the tickler chains. For example, with regard to by-catch, a survey of discards by beam trawlers operating in the Irish Sea between 1993 and 2002 estimated that two-thirds of the catch was discarded (Borges et al. 2005). Groenwold and Fonds (2000) cited literature from 1998 which estimated that the annual discards produced by beam trawling in the southern North Sea amounted to 150–190,000 t of dead fish (mainly dab) and up to 85,000 t of invertebrates. It was noted that the chances of survival for the fish by-catch is low, but higher for crustaceans and molluscs (40–50%) and starfish (90%).

The use of tickler chains in beam trawling inevitably digs out or damages bottom-dwelling sea organisms (Kaiser and Spencer 1996) and can result in significant mortality of non-target species. Jennings et al. (2001) studied the impacts of beam trawling in one region of the North Sea on infauna (animals that live within the sediments) and epifauna (animals that live on the surface of the sediments). Significant decreases in the biomass of bivalves and spatangoids (burrowing sea urchins) were evident. It was suggested that long-term trawling disturbance would lead to dramatic reductions in the biomass of infauna and epifauna. In this regard, it is important to note that some areas of the southern North Sea are trawled on average more than three times per year. However, trawls are not evenly distributed but are patchy, as some areas are more favoured by fishermen than others. It has been estimated that some areas are visited up to 400 times a year while others are never fished (Kaiser et al. 1998).

Other studies have also shown significant decreases in bottom-dwelling animals following beam trawling. For example, beam trawling on the Dutch continental shelf of the North Sea was found to cause mortality of up to 68% of bivalves, up to 49% of crustaceans, and up to 31% of annelid worms (Bergman and van Santbrink 2000). It was noted that those species that live within reach of the tickler chains, and that are not robust, are expected to suffer significant fishing mortalities. Kaiser and Spencer (1996) reported that there was a decrease in the number of some species by 58% after beam trawling and a decrease in diversity (mean number of species per sample) by 50%. These decreases were observed in sediments that were otherwise stable. On mobile sediments, such as those subjected to seasonal storms or strong tidal currents, there are naturally less fauna and no impact of beam trawling could be detected in this type of habitat.

Immediately after an area has been subjected to bottom trawling and many species have been removed or damaged, there may be an increase in other mobile scavenger species. For example the common hermit crab (*Pagurus bernhardus*) and the brittle star (*Ophiura ophiura*) have been shown to increase in trawled areas. These species are scavengers and are known to move quickly into areas of fishing disturbance to feed on animals damaged by trawls (Kaiser et al. 1998). Other changes reported in sea-bed communities following beam trawling included a shift from a community dominated by bivalves and spatangoids to one dominated by polychaete worms (Jennings et al. 2001). However, any increases in the biomass of such smaller species were small in relation to the losses of the overall community biomass that resulted from the depletion of larger individuals (Schratzberger and Jennings 2002).

With regard to the recovery of an area following beam trawling, research appears to be limited. One study reported that, at the locations investigated, there was no difference between trawled and untrawled sites 6 months after an experimental trawl (Kaiser et al. 1998).

Research has been conducted on shell damage and mortality of whelks (*Buccinum undatum*) caught by beam trawling in the southern North Sea (Mensink et al. 2000). Whelks caught in beam trawls had significantly more damage to their shells compared to whelks caught with baited traps. The captured whelks were kept in the laboratory to assess their survival over a period of 6 weeks. Only 40% of whelks caught by a 12-m beam trawl survived, irrespective of the shell damage incurred. By comparison, over 95% of whelks caught with baited traps survived. It was suggested that morality suffered by whelks following beam trawling may be even greater under natural conditions, due to the presence of predators and scavengers. The study implied that the beam trawl fishery could have a major adverse impact on whelk populations and that the increased level of beam trawling in the North Sea over the past 20–25 years might well be the main cause of the declines in whelks in several parts of the North Sea.

2.5.3 Industrial Fishing

The so-called 'industrial fisheries' target fish for conversion to fishmeal or fish oil rather than for direct human consumption. Small pelagic (open water) fish such as anchovies, herrings and sardines are the usual target species of industrial fisheries. Table 2.2 shows the top pelagic species that were caught for reduction to fishmeal and fish oil in 2003. Major uses of fishmeal and fish oil include feedstuffs for farmed fish, livestock and poultry.

Historically, industrial fishing had its beginnings in the 1950s and rapidly expanded in the 1960s with the advent of new technology such as sonar and the Danish or purse seine net (Greenpeace 1997). Other methods of capture include trawling with small mesh nets in the range of 16–32 mm. With regard to the quantity of fish landed by industrial fisheries, the quantity of whole fish reduced to fishmeal and fish oil has stabilised at around 25 million tonnes per year since the beginning of the 1970s, fluctuating between 20 and 30 million tonnes (Tacon et al. 2006).

2.5.3.1 Sustainability of Industrial Fisheries

Industrial fishing has led to the decline and collapse of some fish populations. For example, overfishing led to the collapse of North Sea mackerel (*Scomber scombrus*)

Table 2.2 Top pelagic fish mainly caught for reduction in 2003 (Tacon et al. 2006)

Species	Total reported production (millions of tonnes)	Production by country (%)
Peruvian anchovy (*Engraulis ringens*)	6.2	Peru 86.2%, Chile 13.2%
Blue whiting (*Micromesistius poutassou*)	2.38	Norway 35.7%, Iceland 21%, Russian Federation 15.1%, Faeroe Islands 13.7%, Denmark 3.7%, Sweden 2.7%, The Netherlands 2.4%
Japanese anchovy (*Engraulis japonicus*)	2.09	China 62.3%, Japan 25.6%, Korea Republic 12%
Atlantic herring (*Clupea harengus harengus*)	1.96	Norway 28.7%, Iceland 12.8%, Canada 10.2%, Russian Federation 7.4%, Denmark 5.9%, United States 5%, The Netherlands 4.8%, United Kingdom 4.6%, Sweden 4.4%
Chub mackerel (*Scomber japonicus*)	1.85	Chile 30.9%, China 23.6%, Japan 17.8%, Korea Republic 6.6%, Peru 5.1%
Chilean jack mackerel (*Trachurus murphyi*)	1.73	Chile 81.9%, Peru 12.5%, China 5.4%
Capelin (*Mallotus villosus*)	1.15	Iceland 59.2%, Norway 21.7%, Russian Federation 8.4%, Faeroe Islands 4.4%, Greenland 2.6%, Denmark 1.5%
European pilchard (*Sardina pilchardus*)	1.05	Morocco 62.8%, Algeria 7.3%, Portugal 6.3%
Californian pilchard (*Sardinops sagax*)	0.691	Mexico 89.6%, United States 10.4%
European sprat (*Sprattus sprattus sprattus*)	0.631	Denmark 41.5%, Poland 13.3%, Sweden 12.1%
Gulf menhaden (*Brevoortia patronus*)	0.522	United States 100%
Sandeels (*Ammodytes* spp.)	0.341	Denmark 82.9%, Norway 8.7%, Sweden 6.4%
Atlantic horse mackerel (*Trachurus trachurus*)	0.214	Ireland 21.5%, Norway 9.5%, Germany 8.7%, Portugal 8.7%, Denmark 6.5%, France 5.4%
Norway pout (*Trisopterus esmarkii*)	0.037	Denmark 60.9%, Norway 32.8%, Faeroe Islands 6.2%

in 1970, followed by collapse of herring stocks (*Clupea harengus*) in the mid-1970s. The fishery was forced to turn to smaller species, such as the Norway pout (*Trisopterus esmarkii*) and, later, to the lesser sandeel (*Ammodytes marinus*). Industrial fishing also contributed to the collapse of the Peruvian anchovy (*Engraulis ringens*) in the 1970s. In the 1980s capelin (*Mallotus villosus*) stocks in the Barents Sea and herring (*Clupea harengus*) stocks in the Norwegian Sea collapsed after intense fishing pressure (Greenpeace 1997).

More recently, one report has analysed data on six industrially fished species which are used as feed for finfish aquaculture in Scotland, to determine whether these fisheries are sustainable (Huntington 2004a). These industrially fished species included Peruvian anchovy (*Engraulis ringens*), Chilean jack mackerel (*Trachurus murphyi*), capelin (*Mallotus villosus*), blue whiting (*Micromesistius poutassou*), sandeel (*Ammodytes* spp.) and horse mackerel (*Trachurus trachurus*). The study found that most of the fisheries did not meet requirements of sustainability. For example, it concluded that the Chilean jack mackerel was overfished and is still recovering from previous overfishing; the catch limit on horse mackerel was too high to sustain the fishery; the harvest of blue whiting was considered to be unsustainable; and the sustainability of both capelin and sandeel fisheries was uncertain. There were insufficient data on the Peruvian anchovy to determine whether the fishery was sustainable. However, the species has been subjected to heavy fishing pressure over the years and stock levels are also extremely vulnerable to climatic changes due to the El Niño phenomenon. Currently, stocks are considered to be fully or over-exploited (Tacon 2005; Tacon et al. 2006). In addition to these six species, many other industrially fished species are either fully or over-exploited (Tacon 2005). Given the unsustainable nature of many industrial fisheries, it is of concern that the consumption of fishmeal for use in aquaculture feeds is predicted to increase further, due to expansion of the industry (see Chapter 3, Section 3.3).

2.5.3.2 Ecological Impacts

The fish species targeted by industrial fisheries are low in the food chain, and as such they form a critical base for the marine food web, providing food for marine predators including many commercially valuable fish, marine mammals and seabirds (Naylor and Burke 2005). Consequently, there may be adverse impacts on marine ecosystems and, in particular, for predatory species where there is competition from fishing. Research on the ecosystem effects of overfishing is unfortunately still quite limited. Examples of impacts are given below:

- Fishing induced the collapse of the Norwegian spring-spawning herring (*Clupea harengus*) stock in the late 1960s and the stock has since struggled to recover. Whilst stocks were at their lowest between 1969 and 1987, the breeding success of Atlantic puffins at Røst in the Norwegian Sea was severely impacted due to a reduction in food supply. Fledging success of chicks was less than 50% in all but three seasons and in most years were complete failures (Anker-Nilssen et al. 1997).
- A study on the North Sea sandeel fishery reported a negative impact of the fishery on black-legged kittiwakes (*Rissa tridactyla*) (Frederiksen et al. 2004). These seabirds have declined by over 50% since 1990 and their breeding success and adult survival were low when the Wee Bankie sandeel fishery off Scotland was active. The study also found that increasing sea temperature may also be having a negative impact on sandeel populations, and thus on food supply to the birds. It was recommended that the Wee Bankie sandeel fishery should remain closed indefinitely, as a re-opening would probably lead to a marked decline in numbers of the kittiwakes. On the recommendation of the International Council

for the Exploration of the Sea (ICES) the sandeel fishery east of Scotland was closed from 2000–2005. Sandeel stocks in the whole of the North Sea have since been recognised to be severely depleted, which led the European Commission to close the fishery in July 2005 (EC 2005). In 2006, ICES recommended that the fishery should remain closed until its recovery (ICES 2006).

It has been suggested that the continuing pressure exerted by industrial fisheries at low levels of the food web, combined with an ever-increasing demand for fishmeal by the expanding aquaculture industry, also puts pressure on marine fish predatory species higher up the food chain. Further, it might be difficult for populations of fish occupying higher trophic levels to recover even if pressure on industrially fished species were to be significantly decreased (Deutsch et al. 2007).

2.6   By-catch of Seabirds, Marine Mammals and Turtles

Substantial numbers of marine mammals, turtles and seabirds become entangled or hooked accidentally by fishing gear every year and, consequently, many die. Just as in the entrapment of unwanted fish and invertebrates, this incidental take of other non-target animals is known as by-catch. Fisheries by-catch has been implicated in the declines of many seabird populations (Section 2.6.1) marine mammals (Section 2.6.2) and turtles (Section 2.6.3).

Some action has been taken to alleviate the problems of by-catch. For instance, in December 1992, after a sustained campaign by Greenpeace and others, the UN placed a global moratorium on the use of driftnets greater than 2.5 km in length on the high seas (Hall 1999; Hall et al. 2000). This type of fishing had raised concern because of the huge by-catch of fish, mammals, turtles, sharks and seabirds (Hall 1999). For example, Fig. 2.10 shows a swordfish killed in an illegal driftnet in the Mediterranean, while Fig. 2.11 shows a turtle being rescued from a similar net. However, other types of fishing also result in by-catch and problems continue to the present day.

2.6.1   Seabirds

More than 300,000 seabirds, including 100,000 albatrosses, are estimated to be killed by longline fishing fleets every year (Birdlife International 2005). Longline fishing uses gear consisting of a main line with numerous baited hooks attached on branch lines (Gilman et al. 2005) (see Fig. 2.12). In pelagic longlining, the gear is suspended from a line drifting at the sea surface, and in demersal longlining, the gear is set at the seabed (Gilman et al. 2005). Birds are accidentally killed by both pelagic and demersal longline fisheries because they are attracted to discards and baits and can ingest baited hooks during the setting of the longline or, less

Fig. 2.10 On board the Rainbow Warrior in Italian waters in 2005, an illegal driftnet is cut from a swordfish (Greenpeace/Roger Grace)

Fig. 2.11 Greenpeace crew-members rescue and release a turtle from an illegal Italian driftnet, 2006 (Greenpeace/Roger Grace)

Fig. 2.12 Longliner fisherman
prepares his lines in the port of
Argostoli, on the Greek Island of
Kefalonia, 2006 (Greenpeace/Jeremy
Sutton-Hibbert)

commonly, during its hauling back in. Once caught on a hook, the birds are pulled
underwater by the weight of the line and drowned (Baker and Wise 2005).

Pelagic longline fisheries mainly target tuna, swordfish, billfishes and sharks.
They operate in the temperate seas in the North Pacific and Southern Ocean and
result in a high by-catch of seabirds. Demersal longline fisheries operate in the
North Pacific, North Atlantic and Southern Ocean and target species such as cod,
haddock, tusk, halibut, sablefish and Patagonian toothfish. The fisheries claim a
high by-catch of seabirds (Gilman et al. 2005).

2.6.1.1 Seabird Species Affected by Longline Fisheries

Brothers et al. (1999) reviewed data on seabird by-catch throughout the world
and reported that a total of 61 species of seabirds are known to have been killed
by longline fishing. Of these, 25 (39%) are classified as critically endangered,
endangered, or vulnerable by the World Conservation Union (Brothers et al. 1999).
Mortality from longline fishing is recognised as a significant contributing factor
to the threatened status of these seabirds (Hall and Mainprize 2005; Brothers et
al. 1999). Quantitative data on seabird mortality arising from longline fishing is
limited, although some research does show that populations are at risk.

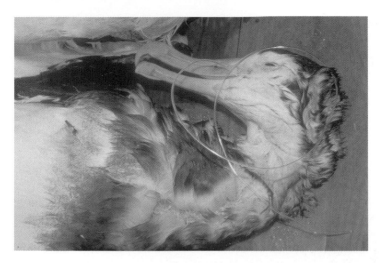

Fig. 2.13 Wandering Albatross caught in Japanese longline in New Zealand waters in 1997 (Greenpeace/Dave Hansford)

Species that are most commonly killed in longline fisheries are albatrosses and larger petrels of the family Procellariidae (FAO 1999) (see Fig. 2.13). This is of particular concern given their slow reproductive rate. While some nations have introduced mitigation methods to help reduce the number of birds caught (see Section 2.6.1.2 below), most longline fleets still do not employ effective mitigation methods (Gilman et al. 2005).

There are 24 species of albatrosses worldwide and 21 species are showing declines in more than 50% of their populations. Most species are classified as threatened. Albatrosses face threats for a number of different reasons, but longlining is known as perhaps the greatest threat (Arnold et al. 2006) and has been implicated in the decline of some populations. Rare and threatened species taken by longlining include:

- The vulnerable short-tailed albatross (*Phoebastria albatrus*) in the North Pacific
- The critically endangered Amsterdam albatross (*Diomedea amsterdamensis*)
- The Chatham albatross (*Thalassarche eremita*)
- The endangered Tristan albatross (*Diomedea dabbenena*)
- The Northern Royal albatross (*Diomedea sanford*) (Brothers et al. 1999)

Declines in the population sizes of the wandering albatross (*Diomedea exulans*) and the southern Buller's albatross (*Thalassarche bulleri bulleri*) have been attributed to mortality caused by longline fisheries in the Southern Ocean according to literature cited by Hyrenbach and Dotson (2003). Research on South Georgia has noted that the wandering albatross had lost nearly one third of its population since 1984. The wandering albatross and the grey-headed (*Thalassarche chrysostoma*) albatross have undergone declines in populations in recent years at sub-Antarctic Marion Island in the southern Indian Ocean. It was suggested that a contributing factor may be recent increases in tuna longlining, as well as recent large-scale Illegal, Unregulated and Unreported (IUU) longline fishing for Patagonian toothfish (*Dissostichus eleginoides*) (Nel et al. 2002). It has been estimated that IUU longline fishing for Patagonian toothfish may take up to 145,000 seabirds annually (Brothers et al. 1999).

A recent study on the black-browed albatross (*Diomedea melanophris*) at South Georgia suggested that the population studied had been declining since 1976 and more rapidly in the last decade, at the same time that longline fishing had intensified (Arnold et al. 2006). Longline fishing was suggested as a primary contributor to the decline, along with a decrease in krill abundance, and it was predicted that the South Georgia population may face extinction this century if the current rate of decline is not abated. As noted above, the wandering albatross and grey-headed albatross have also undergone significant declines on South Georgia. These species will also be faced with extinction without action to abate the problem.

The endangered Tristan albatross (*Diomedea dabbenena*) is known to be killed by longline fisheries in the South Atlantic and a recent study predicted an annual mortality of 471–554 birds, sufficient to cause population decreases of 3.6–4.3% per year (Cuthbert et al. 2005). It was suggested that mitigation methods need to be employed to curtail the threat to these birds.

In the North Pacific, the black-footed albatross (*Phoebastria nigripes*) and Laysan albatross (*Phoebastria immutabilis*) are known to suffer mortalities from longline fishing (Cousins et al. 2000). Estimations suggest that the existence of the black-footed albatross could be threatened by longline fisheries if current mortality rates continue (Gilman et al. 2005).

Brothers et al. (1999) reported that the killing of hundreds or thousands of petrels by the longline fisheries for southern bluefin tuna (*Thunnus maccoyii*) and Patagonian toothfish in the Southern Ocean is also of serious concern. This includes the southern giant petrel (*Macronectes giganteus*), northern giant petrel (*Macronectes halli*), white-chinned petrel (*Procellaria aequinoctialis*) and grey petrel (*Procellaria cinerea*). For example, many thousands of white-chinned petrels are killed off the coasts of northern to subtropical South America and South Africa where they winter (Barnes et al. 1997; Weimerskirch et al. 1999). A recent study showed that breeding adult birds are vulnerable to encounters with longline vessels, particularly from tuna fisheries of the subtropical high seas of the Atlantic and Indian Oceans, from toothfish and hake fisheries off South America and South Africa, and from fishing for toothfish near to their breeding sites at South Georgia, Kerguelen, Crozet and the Prince Edward Islands (Weimerskirch et al. 1999). The study concluded that prospects for this species are bleak unless effective action is taken, and a plea was made for an international global approach for fishing methods that avoid by-catch.

In Eastern Australia, another species of bird, the flesh-footed shearwater (*Puffinus carneipes*) which nests on Lord Howe Island is under threat from longline fishing (Baker and Wise 2005). An estimated 1,794–4,486 shearwaters were killed annually between 1998 and 2002, a rate which threatens the survival of the population.

In the Northern Hemisphere, although long-lining is increasing in European waters, little information is available on seabird mortality, threats to populations and possible mitigation measures. However, one study in the Western Mediterranean around Columbretes Islands (Spain), estimated that 437–1,867 Cory's shearwaters (*Calonectris diomedea*) were killed annually. This is of concern because this species is slow to reach maturity and has a low reproductive rate, such that even small increases in adult mortality may result in population declines (Belda and Sánchez 2001). In the North and Norwegian Seas, longline fisheries of Norway, Iceland and the Faroes are known to take many fulmars (*Fulmarus glacialis*), although data are non-quantitative. The fulmar population is large and longlining is not thought to be a conservation risk, although more information is required (Brothers et al. 1999).

2.6.1.2   Mitigation Methods

A number of mitigation methods for reducing the by-catch of seabirds have been developed. For example, these include:

- Lines of streamers trailed behind vessels over the area where the hooks enter the water to scare birds
- Setting baited lines in total darkness
- Adding weights to longlines to accelerate sink rates
- Setting longlines deep underwater through tubes, thereby setting the lines out of sight of seabirds
- Dyeing baits blue so that birds do not see them as easily (Hall and Mainprize 2005)

Other measures include the establishment of exclusion zones around breeding areas, particularly during certain periods of the breeding season when birds are constrained to remain close to their breeding sites (Cuthbert et al. 2005). The establishment of protected areas around nesting seabird colonies within a nation's Exclusive Economic Zone is one potential way of doing this in coastal areas (Gilman et al. 2005).

Mitigation measures have been adopted by some countries to manage seabird mortality in North Pacific longline fisheries. These include some fisheries in Japan, the USA and Canada, whereas China, Korea, Mexico, Russia and Taiwan do not have regulations to reduce by-catch of seabirds (Gilman et al. 2005). In many Southern Oceanic fleets, the use of mitigation methods is either inconsistent or non-existent (Tuck et al. 2003).

Mitigation methods have proved to be successful in reducing seabird by-catch in a number of fisheries. For example, in the demersal fishery at South Georgia, seabird mortality has been virtually eliminated by not fishing when the birds are breeding. The by-catch of seabirds in a pelagic fishery for swordfish and other tunas off Brazil has been substantially reduced by setting lines at night (Tuck et al. 2003).

Mitigation methods need to be tailored for different fisheries because some methods will be more appropriate than others, depending on the species of seabird in question and other factors concerning the fishing vessel design and fishing method used (Gilman et al. 2005).

The Royal Society for the Protection of Birds and Birdlife International have a 'Save the Albatross' campaign which includes work on educating fishermen in mitigation methods, and lobbying governments to legislate on mitigation methods.

2.6.1.3 National and International Measures to Address Seabird By-catch

In 1998, the Food and Agricultural Organisation of the United Nations (FAO) set up 'The International Plan of Action for reducing the incidental catch of seabirds in longline fisheries (IPOA-SEABIRDS) (FAO 2004b). IPOA-SEABIRDS is voluntary and it aims to encourage states involved in longline fishing to identify where seabird by-catch is a problem and, in so doing, to produce an assessment on how to reduce seabird by-catch effectively – a National Plan of Action (NPOA-SEABIRDS). The NPOA-SEABIRDS should prescribe the appropriate mitigation measures to be taken. By 2007, NPOAs had been adopted by Brazil, Canada, Chinese Taipei, Falkland Islands/Malvinas, Japan, New Zealand, Uruguay and USA. In addition, several other countries had NPOAs that were under development (Gilman et al. 2007).

Regional Fisheries Management Organisations (RFMOs) are inter-governmental organisations that have the responsibility of managing the high seas and migratory fish stocks such as tunas, swordfish, cod, toothfish and billfish. There are 19 RFMOs, of which 16 are active (Birdlife International 2005). RFMOs are legally required to manage not only the target fish species but also the overall ecosystem. A recent report by the NGO Birdlife International reported that only 1 of the 19 RFMOs (the Commission for the Conservation of Antarctic Marine Living Resources, CCAMLR) had taken comprehensive action to mitigate seabird by-catch (Small 2005). By-catch had been substantially reduced under CCAMLR in the conventional area, from 6,589 birds killed in 1997 to 15 birds killed in 2003. However, such reductions had not been achieved by other RFMOs. For example, the report highlighted five RFMOs that had the greatest regional overlap with albatross distribution. Only CCAMLR scored well in the assessment. The Western and Central Pacific Fisheries Commission (WCPFC) had made commitments to reducing by-catch but it was not possible to assess their progress because the RFMO was not fully active at the time of the assessment. The Commission for the Conservation of Southern Bluefin Tuna (CCSBT) had implemented one mitigation measure but scored poorly because this was insufficient. The remaining two RFMOs, the Indian Ocean Tuna Commission (IOTC) and the International Commission for the Conservation of Atlantic Tunas (ICCAT), had not implemented any mitigation measures. A more recent assessment of action taken on seabird by-catch was undertaken by the FAO (Gilman et al. 2007). The report listed several RFMOs that have taken steps to address seabird interactions in marine capture fisheries, including the CCAMLR, CCSBT, IOTC, Inter-American Tropical Tuna Commission (IATTC), International Commission for the Conservation of Atlantic Tunas (ICCAT),

International Pacific Halibut Commission (IPHC), South East Atlantic Fisheries Organisation (SEAFO) and WCPFC.

It has been stated that Illegal, Unreported and Unregulated (IUU), or pirate, fishing (see also Chapter 6, Section 6.1) is responsible for about one third of the annual seabird deaths caused by longlining (Birdlife International 2005). Reducing IUU fishing, and the strong control of legal fleets through the enforcement of appropriate mitigation measures have been suggested (Cuthbert et al. 2005). Greenpeace is demanding that governments take action to prohibit access for pirate vessels to ports, close markets to their produce and prosecute companies supporting them (Bours et al. 2001).

Gilman et al. (2005) noted that the potential exists to minimise seabird mortality in longline fisheries to insignificant levels. It was suggested that, as a global and multinational problem, a significant reduction in seabird by-catch requires international collaboration by management authorities, industry and other stakeholders to share technical and financial resources to mitigate seabird and longline interactions.

2.6.2 Marine Mammals

Fishery operations can cause the mortality of, or serious injury to, marine mammals that are 'captured' but discarded. Dolphins and harbour porpoises are the most common casualties among cetaceans (whales, dolphins and porpoises), and seals and sea lions (pinnipeds) are also affected (Read et al. 2006). When these small marine mammals are caught in fishing gear, they often die because they are not strong enough to break free and come to the surface to breathe. Larger whales may also become entangled in fishing gear but usually they can break free, although they may continue to tow gear for long periods. This can cause injuries and even slow death if the gear interferes with feeding (CBRC 2006).

Several types of fishing methods can lead to marine mammal by-catch, including purse-seine nets, gillnets, trawl nets and driftnets (Hall et al. 2000) (see, for example, Fig. 2.14). Gillnets are in widespread use and are suspected to be responsible for a high proportion of global marine mammal by-catch. For example, research in the USA revealed that the majority of cetacean (84%) and pinniped (98%) by-catch between 1990 and 1999 occurred in gillnet fisheries (Read et al. 2006).

Presently, fisheries are posing the single greatest threat to many populations of marine mammals (Read et al. 2006). In 2002, a meeting of a group of cetacean experts concluded that incidental capture in fishing operations is a major threat to whales, dolphins and porpoises worldwide. Several species and many populations will be lost in the next few decades if nothing is done. However, despite the global nature of this problem, awareness of the issue if generally low (Read and Rosenberg 2002).

One study has estimated the global by-catch of marine mammals by extrapolating data from research on marine mammal by-catch on US vessels in the 1990s. The estimate for the annual global by-catch of cetaceans was staggering, at 307,753 animals.

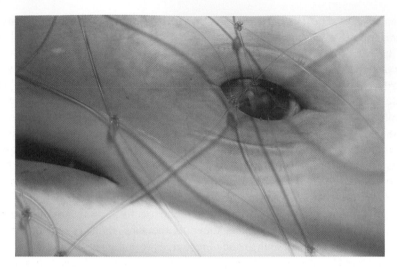

Fig. 2.14 A Pacific white-sided dolphin (*Lagenorhynchus obliquidens*) caught in a driftnet in the North Pacific in 1990 (Greenpeace/Roger Grace)

The estimate for pinnipeds was even higher at 345,611 animals (Read et al. 2006). The study suggested that the estimate for pinnipeds is likely to be positively biased because there is little by-catch in many tropical countries. However, the estimate for cetaceans is likely to be an underestimate because the registry of fishing vessels in the FAO database on which figures were based is incomplete.

Research on by-catch of marine mammals in fisheries worldwide is generally fragmentary. In particular, a search of the scientific literature on seals and sea lions over the past 10 years yielded few studies, and problems with by-catch of these animals may therefore go unreported. Studies on cetaceans indicated that there is cause for concern for many species (see below, Sections 2.6.2.1–2.6.2.3). By-catch is the main threat to several endangered species of marine mammals including the vaquita (*Phocoena sinus*) and Hector's dolphin (*Cephalorhynchus hectori*), and contributes to the poor conservation status of the Mediterranean monk seal (*Monachus monachus*) and the right whale (*Eubalena glacialis*) in the North Atlantic (Read et al. 2006).

2.6.2.1 Dolphins

In the 1950s, a new method of catching tuna using purse-seine nets was implemented in the eastern Pacific Ocean. The fishery resulted in the deaths of what were estimated to be hundreds of thousands of dolphins in the 1960s until pressure from environmental groups and public outcry led to the passage of the Marine Mammal Protection Act in 1972 (Hall et al. 2000; Lewison et al. 2004a). Further pressure came from environmental groups, which led to trade restrictions on the sale of tuna captured with a by-catch of dolphins. In response, the tuna fishery developed innovative mitigation methods which reduced dolphin mortality and resulted in most dolphins being released alive (Lewison et al. 2004a). Several changes in fishing gear, changes in setting of the nets, and hand rescue by divers were implemented as solutions to the problem. Mortalities of dolphins in 1986 from the fishery were estimated at 133,000 but had been reduced to 1,877 by 1998. Most of the dolphin populations had declined by the early 1970s and several stocks were classified as depleted in the early 1990s (Hall et al. 2000). Although recent levels of mortality are no longer significant from a population point of view (Hall et al. 2000), dolphin populations have not yet recovered (Lewison et al. 2004a). The continual sub-lethal effects of prolonged chasing of the dolphins by the fishery when hunting tuna and their frequent capture may contribute to a reduction in breeding success and their failure to recover (Lewison et al. 2004a).

Hector's dolphin (*Cephalorhynchus hectori*) is endemic to the coastal waters of New Zealand. Incidental entanglements in gillnets have led to population declines and the species was classified as endangered by the IUCN in 2000 (Reeves et al. 2003). There are four separate populations of Hector's dolphin around New Zealand and in the North Island population there are only 111 individuals (Slooten et al. 2006). This population is known as Maui's dolphin (*Cephalorhynchus hectori maui*) and has been classified by the IUCN as critically endangered (Reeves et al. 2003). A protection area was established in 2003 in an attempt to halt the decline of Maui's dolphin and gillnetting was prohibited in this zone. However, gillnetting continues in harbours, and trawling continues in the protected area, as does illegal gillnetting. The continued finding of Maui dolphin carcasses bearing scars from fishing nets is evidence that by-catch is still occurring. Research suggests that by-catch needs to be eliminated completely to allow the population to recover and, consequently, action is needed to prevent gillnetting in harbours, in the protected area and in adjacent areas (Slooten et al. 2006).

Short-beaked common dolphins (*Delphinus delphis*) were abundant in the northern part of the Western Mediterranean up to the 1970s but are now rarely sighted there. Reeves et al. (2003) suggested that at least some of the decline in numbers is due to illegal driftnet fishing operations by Spain, Italy and Morocco. Figures obtained for the Spanish fleet in the early 1990s suggested that the by-catch of short-beaked common dolphins and striped dolphins (*Stenella coeruleoalba*) in the region were possibly not sustainable (Silvani et al. 1999). Also of concern for the short-beaked common dolphin in Atlantic waters off western Europe was large-scale mortality in trawl nets, tuna drift nets and sink gillnets, reported in the 1990s (Reeves et al. 2003).

Reeves et al. (2003) noted that recent by-catch of dusky dolphins (*Lagenorhynchus obscurus*) off Peru in gillnet fisheries was cause for serious concern. A study of dusky dolphins in the Patagonian waters of Argentina, reported that the Patagonian trawling fishery caused the incidental mortality of the dolphins and could be a threat to the population (Dans et al. 2003).

Hundreds of northern right whale dolphins (*Lissodelphis borealis*) are reported to be killed each year in gillnets used to catch billfish, sharks, squid and tuna inside the Exclusive Economic Zones (EEZ) of North Pacific countries. There is also concern that large numbers of southern right whale dolphins (*Lissodelphis peronii*) are taken as by-catch in the driftnet fishery for swordfish that began in northern Chile in the early 1980s. In the Gulf of Thailand, there is concern regarding the by-catch of dwarf spinner dolphins (*Stenella longirostris longirostris*), for which there is no catch monitoring programme or population assessment programme. The Irrawaddy dolphin (*Orcaella brevirostris*) has been severely depleted in parts of Thailand and by-catch in fisheries (e.g. gillnets, explosives) is most likely to be the main cause of depletion. In Taiwan, the incidental take of the Indo-Pacific bottlenose dolphin (*Tursiops aduncus*) in gillnets has been identified as being of concern (Reeves et al. 2003).

2.6.2.2 Porpoises

The harbour porpoise (*Phocoena phocoena*) is widely distributed in coastal waters of the temperate and sub-arctic Northern Hemisphere. Their numbers have been seriously impacted by fisheries by-catch in some areas. Although not threatened as a species, the reduction of numbers in some areas has left regional populations that are severely depleted. For example, populations in the Baltic Sea, Black Sea and Sea of Azov have been classified as vulnerable by the IUCN (Reeves et al. 2003). In the North Sea, estimates suggest that several thousand are killed each year in nets (Hall et al. 2000). A study in the Gulf of Maine, USA, and the Bay of Fundy, Canada, concluded that incidental catch by fisheries was a threat to harbour porpoise populations (Caswell et al. 1998).

The vaquita (*Phocoena sinus*) is a porpoise which inhabits waters of the upper Gulf of California, Mexico. A survey from 1997 suggested that population numbered only 567 individuals and the species has been listed by the IUCN as being critically endangered (Rojas-Bracho et al. 2006). The principle threat to the vaquita is a high by-catch rate in gillnet fisheries (D'Agrosa et al. 2000). Government agencies in Mexico have acknowledged the need to confront the problem of by-catch if the vaquita is to be saved from extinction. However, according to Rojas-Bracho et al. (2006), progress on preventing by-catch has been slow despite efforts to phase out gillnets in the vaquita's core range and the development of schemes to compensate fishermen. Fortunately, at the end of 2005, the Mexican Ministry of Environment declared a Vaquita Refuge covering, within its borders, about 80% of legitimate vaquita sightings. Rojas-Bracho et al. (2006) noted that its effectiveness remains to be seen.

High numbers of Dall's porpoises (*Phocoenoides dalli*) are known to be killed in drift nets within the national waters of Japan and Russia where the UN ban on drift nets does not apply. Between 1993 and 1999, 12,000 were estimated to die as by-catch in the Japanese salmon driftnet fishery operating in the Russian EEZ (Reeves et al. 2003).

2.6.2.3 Whales

Right whales (*Eubalaena* spp.) in the Northern hemisphere face a serious threat from both ship-strikes and entanglement with fishing gear (Reeves et al. 2003). The

North Atlantic right whale (*Eubalaena glacialis*) is particularly at risk because it is one of the most endangered whales in the world and inhabits regions that are heavily used by shipping and the fishing industry (Kraus et al. 2005). The population is estimated to number only 350 animals. Since 1986 there have been 50 reported deaths and at least six of these were due to fishing gear entanglements. During the same period, there were 61 confirmed cases of entanglement, with 12 deaths suspected. Entangled whales usually sink after death so these figures are likely to be an underestimate. Calculations suggest that this species is well on its way to extinction and measures are therefore needed urgently to stop mortality as a result of collisions with ships and entanglement with fishing gear (Caswell et al. 1999). Some action has been taken to close selective areas to fishing and to change fishing gear, but these have not been effective and other changes to fishing gear and fishing methods have been proposed in a recent study (Kraus et al. 2005). In addition, an emergency measure to reduce speeds and re-route shipping has been suggested, to prevent collisions.

Humpback whales (*Megaptera novaengliae*) of the North Atlantic are also endangered and suffer a significant problem from entanglement in fishing gear (Johnson et al. 2005). The majority of entanglement has been found to be due to pot and gillnet gear. In the Gulf of Maine, research showed that more than half the population had scars indicating previous entanglement and 8–25% incurred new injuries each year. In another study, of 30 humpback whales that had become entangled, three were known to have died as a result (Johnson et al. 2005).

Beaked whales (*Mesoplodon* spp.) are deep-water animals that inhabit waters from cold, temperate and sub-polar latitudes to the tropics. Entanglement in fishing gear, in particular gillnets in deep water, is likely to be the most significant threat to these whales (Reeves et al. 2003). Cuvier's beaked whale (*Ziphius cavirostris*) is known to be killed by fishing gear in many areas including the Mediterranean, Sri Lanka, Taiwan and the west coast of North America. It has been suggested that by-catch of these whales is probably sufficient to warrant a conservation concern (Reeves et al. 2003).

The common minke whale (*Balaenoptera acutorostrata*) is subject to mortality in fishing nets and pots, although this has been given little attention. Of further concern is that the meat and blubber of these animals have commercial value in Japan and the Republic of Korea. There is an incentive to set gear intentionally in areas where a catch of minke whales is likely, or even to 'drive' whales towards the nets. This has been a source of controversy within the Scientific Committee of the International Whaling Commission (Reeves et al. 2003).

2.6.2.4 Mitigation Methods

The above discussion clearly indicates that incidental capture or entanglement in fishing gear is a major threat to some marine mammals. The trend is increasing and is likely to continue, given the increasing industrialisation of fisheries and expansion into new areas such as the high seas (Read et al. 2006). Given the seriousness and widespread extent of the threat, there is an urgent need to implement mitigation methods to deal with the problem. The suitability of mitigation methods should be determined for each species in each fishery since a mitigation measure that works for one species in a particular fishery may not work for another (Ross and Isaac 2004). Mitigation methods include:

- The use of acoustic alarms or 'pingers' which alert cetaceans to the presence of fishing gear and/or stimulate them to swim away
- The use of weights at the top of fishing nets that allow small marine mammals to swim over and away from the nets
- The temporary closure of fishing grounds in cases where marine mammals migrate in and out of fishing areas
- Releasing cetaceans alive from fishing gear (Read and Rosenberg 2002; CBRC 2006)
- The use of exclusion devices (metal grids) within trawl nets (Ross and Isaac 2004)

These mitigation methods have met with some success. For example, in the Gulf of Maine, US, a combination of mitigation measures including the use of acoustic alarms and time-area closures of fishing grounds substantially reduced by-catch of harbour porpoises (*Phocoena phocoena*), from 2,900 in 1990 to 332 in 1999 (Read et al. 2006). By-catch mitigation methods were used during the late 1990s in several US fisheries and a 40% reduction in marine mammal by-catch was recorded between 1990 and 1999, although some of the reduction came from a reduced fishing effort (Read et al. 2006). Acoustic alarms have been shown to be effective for harbour porpoises in the North Atlantic and North Pacific gillnet fisheries, and for common dolphins (*Delphinus* spp.) in a drift net fishery in the North Pacific. Effectiveness has not been proved with other species. Time-area closures reduced by-catch of Hector's dolphins (*Cephalorhynchus hectori*) in New Zealand. Releasing cetaceans whilst alive from fishing gear has been effective with dolphins in purse seine fisheries for yellowfin tuna, with harbour porpoises (*Phocoena phocoena*), humpback whales (*Megaptera novaeangliae*), minke whales (*Balaenoptera acutorostrata*) and North Atlantic right whales (*Eubalaena glacialis*) in herring weirs in the Bay of Fundy, and with humpback whales in a variety of coastal fisheries in the North West Atlantic (Read and Rosenberg 2002).

There is some concern among conservationists regarding the use of acoustic alarms. It has been recommended that alarms should be tested in each area they are to be used, to monitor their efficiency. In addition, research shows that marine mammals may not leave an area because a loud sound is introduced to them once they become used to the sound – a phenomenon known as habituation. It is also important to realise that alarms may drive the animals out of their preferred areas such as feeding and breeding grounds, with the potential for detrimental effects in the longer term (Ross and Issac 2004).

Mitigation measures have also been introduced to reduce by-catch of seals and sea lions. For example, in the krill-trawling fishery around South Georgia, mitigation methods including gear modifications have been employed to avoid fur seal

(*Arctocephalus gazella*) deaths. In 2004, mitigation was reported to be successful and either eliminated seal deaths or greatly reduced them (Hooper et al. 2005).

2.6.2.5 National and International Measures to Address Marine Mammal By-catch

Some attempts have been made by national governments and international organisations to address the problem of marine mammal by-catch. These include:

- The US Marine Mammal Act mandates that incidental mortality of marine mammals should be biologically sustainable and that there should be decreases to levels approaching zero (Hall and Mainprize 2005; NOAA 2007). As a result, in US waters significant steps towards reducing by-catch have been made (Ross and Isacc 2004).
- In 2004, the EC laid down measures on by-catch of cetaceans (EC 2004). This regulation specified a restriction on the use of driftnets by January 2008, necessitated certain vessels to use acoustic pingers and required monitoring of the incidental capture and killing of cetaceans, together with further research and conservation measures to ensure that by-catch of cetaceans does not have a significant impact on the species concerned.
- The Agreement for the International Dolphin Conservation Program for the eastern Pacific has 11 signatory countries and has implemented management to reduce dolphin by-catch (Read and Rosenberg 2002).
- The Agreement on the Conservation of Small Cetaceans of the Baltic and North Seas (ASCOBANS) has defined unacceptable levels of the incidental mortality of small cetaceans in commercial fisheries. Some progress has been made with respect to harbour porpoise by-catch in the Baltic Sea (Ross and Isacc 2004).
- The Scientific Committee of the International Whaling Commission has investigated the scale of cetacean by-catches and reviewed mitigation methods (Read and Rosenberg 2002).

The Cetacean By-catch Resource Centre is supported by the WWF and was established following a meeting in 2002 of a group of experts from around the world on cetacean by-catch. It aims to assist fishermen, scientists, environmentalists and the public, working together to address cetacean by-catch. It has formulated a draft international strategy for reducing the incidental mortality of cetaceans in fisheries (Read and Rosenberg 2002).

2.6.3 Marine Turtles

Of the seven species of sea turtles recognised worldwide, six are listed in the IUCN Red List of Threatened Species (IUCN 2006). The green turtle (*Chelonia mydas*), loggerhead turtle (*Caretta caretta*) and olive ridley turtle (*Lepidochelys olivacea*) are listed as endangered, and the leatherback turtle (*Dermochelys coriacea*), hawksbill turtle (*Eretmochely imbricate*) and Kemp's ridley turtle (*Lepidochelys kempii*) are listed as critically endangered. By-catch of sea turtles in fisheries is a considerable threat to some of their populations. Otter trawls, gillnets and long lines are known to take significant numbers of sea turtles (Alverson et al. 1994).

Turtles breathe air and, when caught by a trawl net, they may lapse into a coma and drown. In addition, if they are caught in nets but do not drown or become comatose, their physical condition may be weakened, putting them at greater risk from predation (Robins 1995). Turtle by-catch from pelagic longlining is a result of turtles attempting to swallow bait on hooks or becoming entangled in gear (Lewison et al. 2004a).

2.6.3.1   Trawl Fisheries

Trawl fisheries for shrimp in the Gulf of Mexico, northern Australia and Orissa, on the east coast of India, are known to be a major cause of turtle mortality due to by-catch (Pinedo and Polacheck 2004). For instance, it has been estimated that as many as 50,000 loggerheads and 5,000 Kemp's ridley sea turtles drowned each year in the 1980s in the southeastern USA and Gulf of Mexico fishery (Alverson et al. 1994). As a consequence, the National Marine Fisheries Service of the US federal government worked with the commercial shrimp trawl industry to develop Turtle Excluder Devices (TEDS). A TED is a metal grid of bars with an opening, which is fitted at the top or bottom of a trawl net. While small animals like shrimp slip through the bars to be caught in the bag end of the net, large animals such as turtles and sharks strike the bars and are ejected through the opening in the net (Lewison et al. 2003). TEDS were required to be fitted into shrimp trawl nets on US vessels by 1991. Because sea turtles are slow to reach reproductive maturity, it may take decades to see the effects of implementing TEDS in fisheries (Alverson et al. 1994).

Although TEDS have been a successful way of releasing turtles unharmed, unfortunately not all fishermen comply with the US law on TEDS and sea turtles continue to drown in shrimp nets (CCC/STSL 2003). In the Gulf of Mexico, stranding rates of turtles in recent years have continued to increase despite the implementation of TEDS. Strandings, where turtles are found immobile, injured or dead, are considered to be an indicator of the effects of commercial fisheries. Research showed that higher levels of strandings were linked to both increased human population levels and an increase in shrimping activity. Results also suggested that TED regulations may have led to a modest decline in the proportion of stranded turtles that die. However, where there was a lack of compliance with TED regulations, this was correlated with a higher level of strandings. In conclusion, the study indicated that improved compliance with TED regulations will reduce strandings to levels that should aid recoveries for the loggerhead turtle and Kemp's ridley turtle. Local seasonal fishery closures, in addition to TEDS enforcement, could reduce strandings to even lower levels (Lewison et al. 2003).

Other research has shown that:

- On the east coast of Australia, in Queensland, otter trawling was found to catch loggerhead turtles, green turtles and flatback turtles (*Natator depressus*) (Robins 1995). About 1.1% of turtles were dead when landed. Another 5.7% of turtles were comatose and potentially could die if not resuscitated by fishermen.
- A study of trawling by Italian vessels in the northern Adriatic reported that 9.4% of loggerhead turtles were landed dead, and potential mortality, assuming all comatose turtles would die, was 43.8% (Casale et al. 2004). It was concluded that trawling in the northern Adriatic is likely to represent a serious threat to turtle populations. It was suggested that the use of TEDS was not a realistic solution for the type of fishing gear used by this fishery but that time-area closures would be likely to help. As an urgent measure it was suggested that an awareness campaign among captains of vessels was needed, to provide information on appropriate recovery procedures for comatose turtles.
- In India, olive ridley turtles nest on the beaches of Orissa. They are threatened by trawling in the region and Greenpeace is campaigning to protect their nesting sites as well as carrying out an education campaign with local people and fishermen (Greenpeace 2006a).

2.6.3.2 Pelagic Longlining

Pelagic longlining is used to catch tuna and swordfish and has been identified as being a major cause of sea turtle by-catch. Fisheries known to be involved include those in the western North Atlantic, the temperate western Pacific, the Mediterranean, the Gulf of Mexico, the Caribbean, the Azores, Madeira, Costa Rica, Hawaii and the western South Atlantic (Pinedo and Polacheck 2004). A recent study estimated that, worldwide, the by-catch of turtles by pelagic longlining in 2000 resulted in the incidental take of over 200,000 loggerhead and 50,000 leatherback sea turtles (Lewison et al. 2004b). Figure 2.15 shows a turtle caught by a Korean longliner in the Pacific near Kiribati.

In the Pacific, there have been population declines of 80–95% of loggerhead and leatherback turtles in the past 20 years, and Lewison et al. (2004b) warned that continued unmitigated longlining would have serious consequences for these turtle populations. It was suggested that multinational efforts are urgently needed to develop and implement mitigation measures to reduce or eliminate turtle by-catch globally. In the Atlantic, the US longline fleet was estimated to have caught between 800 to 3,000 turtles between 1992 and 2000 and this led the National Marine Fisheries Service of the US federal government to issue the Biological Opinion that pelagic longlining in the Atlantic jeopardises the existence of both the loggerhead and leatherback turtles.

In Hawaii, the pelagic longline fishery has been constrained by court orders that have implemented time-area closures of the fishery, primarily as a result of the by-catch of leatherback turtles (Pinedo and Polcheck 2004). In the western

Fig. 2.15 A turtle caught on the end of a bait line of a Korean longliner in the Pacific waters of Kiribati in 2006 (Greenpeace/Alex Hofford)

Mediterranean, studies have suggested that the annual mortality caused by longlining is about 7,500 turtles (Carreras et al. 2004).

2.6.3.3   Other Fishing Methods

Other fishing methods have also been found to result in a by-catch of turtles. In the fisheries off the Balearic Islands in the western Mediterranean, lobster trammel nets from locally based boats were shown to be a major cause of incidental catch of loggerhead turtles in the region (Carreras et al. 2004). Results suggested that a few thousand turtles would be killed annually by lobster trammel nets in the whole western Mediterranean. On the Pacific coast of Mexico at Bahía Magdalena, there are large numbers of green, loggerhead and olive ridley turtles. The area is also important for artisanal fisheries, with bottom set gillnets the most commonly used method. Turtles are frequently caught by this fishing gear, both intentionally for consumption (even though they are nationally and internationally protected), and incidentally. This results in high rates of mortality of turtles in the region (Koch et al. 2006).

2.6.3.4   Mitigation Methods

In the shrimp trawling industry, TED programmes comparable to the programme set up in the USA have been implemented in 15 other countries which export shrimp to the USA. This work has been conducted by NOAA Fisheries (National Oceanic and Atmospheric Administration, US Department of Commerce) and the

US Department of State (NOAA Fisheries 2006a). The promotion of TEDS in other countries is also carried out by the Inter-American Convention for the Protection and Conservation of Sea Turtles, an inter-governmental treaty which provides the legal framework for countries in the American Continent to take actions to benefit sea turtles (IAC 2006). TED programs have been cited as a 'success story' of fisheries by-catch mitigation. However, non-compliance with TED programmes still occurs. For instance, non-compliance was found to be apparent in shrimp fisheries in the Gulf of Mexico (Lewison et al. 2003).

Mitigation methods for pelagic longlining which have been shown to be successful on an experimental basis include the use of circular hooks and mackerel bait. In the North Atlantic, these measures significantly reduced turtle by-catch and reduced turtle mortality in those that were caught, without altering the catch rate of target species (Watson et al. 2005). It was suggested that these methods could be exported to other fisheries and countries for evaluation, but it was also noted that more research worldwide is needed to develop effective by-catch mitigation techniques for longline fisheries. Lewison et al. (2004b) suggested that multi-national efforts are needed immediately to implement successful mitigation measures and to continue to develop measures so that turtle by-catch can be reduced or eliminated.

A recent study on loggerhead turtles that began their life at Cape Verde, West Africa, used satellite tracking to monitor movements of the turtles (Hawkes et al. 2006). It was found that larger individuals foraged in coastal areas whereas smaller individuals foraged in deeper waters, and that all travelled for very long distances. This has implications for their conservation. The long distances travelled by the turtles necessitate international cooperative efforts across seven African states. In addition, turtles forage in international waters including the eastern central Atlantic, which is a hotspot for pelagic longlining. Conservation in these waters requires international action. It was suggested that conservation priorities should focus on less harmful gear types and increased coverage of observer and training programmes.

2.6.3.5 National and International Measures to Address Turtle By-catch

The implementation of TEDS in shrimp fisheries has been discussed above. For other fisheries, Lewison et al. (2004b) suggested that the actions of individual nations alone will not be sufficient to prevent sea turtles from becoming extinct, due to fishery interactions and the fact that the conservation of turtles will require ocean-scale assessments in conjunction with international action.

An international expert workshop on marine turtle by-catch, attended by 19 countries and four inter-governmental organisations (including the FAO) in 2003, made a number of suggestions with regard to regulatory approaches to help the problem of turtle by-catch (International Technical Expert Workshop on Marine Turtle By-catch in Longline Fisheries 2003). Suggestions included:

• Developing new approaches to time-area closures of fisheries
• Encouraging the dissemination of information on the minimisation of turtle by-catch through RFMOs or other regional bodies

• An implementing regulation requiring minimisation of turtle by-catch, based on available research and information on gear modification

Subsequently, in 2004, the FAO convened a technical consultation on sea turtle by-catch and produced 'Guidelines to Reduce Sea Turtle Mortality in Fishing Operations' (FAO 2005b). The guidelines were endorsed at the 26th session of the FAO Committee on Fisheries in 2005 and immediate implementation by Members and regional fishing bodies was called for. In 2007, the FAO reviewed the progress by Members and RFMOs that had been made to address sea turtle by-catch (Gilman et al. 2007; FAO 2007b). Although some progress had been made, formal commitment to, and actual implementation of, the FAO guidelines was not yet standard in the relevant fisheries commissions.

In addition to FAO guidelines, there are also three multilateral agreements with the primary responsibility of regional sea turtle conservation. These agreements are the Inter-American Convention for the Protection and Conservation of Sea Turtles, the Memorandum of Understanding on the Conservation and Management of Marine Turtles and their Habitats of the Indian Ocean and South-East Asia, and the Memorandum of Understanding Concerning Conservation Measures for Marine Turtles of the Atlantic Coast of Africa. These agreements include sea turtle conservation and protection issues, and incorporate provisions to address interactions with fisheries. The agreements do not have fisheries management authority, but they do oblige member States to take by-catch related actions for the areas under their jurisdiction (Gilman et al. 2007).

2.7   Towards Sustainable Fisheries

In simple terms, a particular seafood is sustainable if it comes from a fishery whose practices can be maintained indefinitely without reducing the target species' ability to maintain its population, and without adversely impacting on other species within the ecosystem directly, by removing their food source or by damaging their physical environment (Dorey 2005). On the basis of these basic criteria, most current world fisheries cannot be considered to be sustainable.

This chapter has revealed that current fisheries management has failed to manage fisheries in a sustainable way. Many fish stocks have collapsed in recent years and stocks of many predatory species have been severely depleted. The highly destructive fishing practice of bottom trawling has been permitted to continue. By-catch of seabirds, marine mammals and turtles is of serious concern.

Fisheries management has been based upon the management of singles species. The major considerations are the target species and the non-target species (by-catch). Limits imposed have tended to be in the form of catch quotas, temporarily closed areas and limits upon fishing effort. However, on this basis fisheries management has generally fallen short of the protection of marine ecosystems. In recent years, the concept of ecosystem-based management of marine systems has been formulated, wherein the starting point is the whole ecosystem, not a stock assessment. Definitions of an Ecosystem-Based Approach to Fisheries (EAF) are varied but generally all stipulate that the extraction of resources from marine ecosystems should be sustainable and encompass a precautionary paradigm. However, there are problems in the development of EAF because current understanding of the structure and function of ecosystems is not enough to permit effective, responsive, responsible management of ecosystems as a whole. One approach does exist, which sidesteps this problem and, importantly, will help remediate the considerable degradation currently suffered by marine ecosystems: this is the concept of establishing marine reserves (see also Chapter 7).

Marine reserves can be defined as areas of the ocean completely protected from all extractive and destructive activities. Almost all the marine reserves that have been established to date are small-scale and coastal. They have been shown to benefit biodiversity, leading to increases in biomass, size of individuals and diversity (Halpern 2003) and have also been shown to benefit fisheries in surrounding waters. Evidence of the benefits of marine reserves from research helped to form the basis for a consensus statement published at a Symposium held at the 2001 Annual meeting of the American Association for the Advancement of the Sciences (AAAS). Benefits both inside and outside of reserves were highlighted and it was concluded that existing scientific information justifies the immediate application of fully protected marine reserves as a central management tool for marine ecosystems. It was also stated that there is the need for networks of marine reserves spanning large geographical distances (American Association for the Advancement of Science 2001).

Other bodies have also recognised the need for large areas and networks of marine reserves to be established. For example, the UK Royal Commission on Environmental Pollution in their 2004 report on fisheries (RCEP 2004) suggested that:

> Selection criteria should be developed for establishing a network of marine protected areas so that, within the next five years, a large scale, ecologically coherent network of marine protected areas is implemented within the UK. This should lead to 30% of the UK's Exclusive Economic Zone being established as no-take reserves closed to commercial fishing.

Other bodies have also set goals for establishing marine reserves (see Chapter 7, Section 7.5). Greenpeace believes that it is undoubtedly necessary to set aside large areas as marine reserves, covering 40% or more of marine areas globally, in order to achieve both the conservation goals and accrue the desired benefits to fisheries. The goal will not only be the establishment of marine reserves, but also the sustainable use of marine resources outside the marine reserve network. This implies that these activities must conform to the principles of sustainability, causing no degradation of ecosystem structure and function, and also meet the needs of both current and future generations.

Chapter 3
Aquaculture

Abstract The farming of aquatic plants and animals is known as aquaculture. The production of fish, crustaceans and shellfish by aquaculture has become the fastest-growing animal food sector in the world. Today, aquaculture supplies an estimated 43% of all fish that is consumed by humans globally.

Species that dominate world aquaculture are those at the lower end of the food chain, that is, aquatic plants, shellfish, herbivorous fish (plant-eating) and omnivorous fish (eating both plants and animals). However, marine aquaculture of carnivorous (animal eating) species is also increasing, most notably salmon and shrimp and, more recently, other marine finfish.

The growth of commercial aquaculture has brought with it more intensified methods of production. In some instances, particularly for carnivorous species, intensive methods of aquaculture have created serious environmental problems. There have also been human rights abuses associated with commercial aquaculture in a number of countries. The environmental and negative human rights impacts associated with intensive methods of aquaculture production are discussed in this chapter, using the examples of salmon farming and shrimp farming.

Of further concern is the growing dependence of aquaculture on fishmeal, fish oil or low value fish as feed. Many of the so-called 'industrial fisheries' that are used to supply fish for reduction to fishmeal and fish oil are unsustainable. In addition, the production of carnivorous species by aquaculture requires more wild-caught fish as feed, by weight, than the overall output of the farmed fish. Hence there is a net loss of fish protein in the farming of carnivorous species. In cases where the wild-caught fish used as feed can also be consumed by humans, the issue of food security is raised.

Much work has been done on trying to replace fish-based feeds with plant-based feeds. For some herbivorous and omnivorous fish, it has been possible to replace completely any fishmeal in the diet with plant-based feedstuffs, without impacting on fish growth. Therefore, cultivating such species in this way suggests a more sustainable future path for aquaculture so long as the plant feedstuffs are derived from sustainable sources. However, for carnivorous finfish, it has not been possible so far to replace fishmeal and fish oil completely in the diet.

In order for aquaculture operations to move towards sustainable production, the industry needs to recognise and address the full spectrum of environmental and societal impacts caused by its operations. Essentially, this means that it will no

M. Allsopp et al., *State of the World's Oceans,*
© Springer Science+Business Media B.V. 2009

longer be acceptable for the industry to place burdens of production (such as the disposal of waste) onto the wider environment. In turn, this implies moving towards closed production systems. A series of recommendations by Greenpeace are given towards promoting aquaculture that is both sustainable and fair.

Keywords fish farming, shrimp, salmon, eutrophication, mangrove, habitat destruction, disease, escapes, chemical inputs, salinisation, fish meal, human rights, sustainable aquaculture.

3.1   Introduction

The farming of aquatic plants and animals is known as aquaculture and has been practised for around 4,000 years in some regions of the world (Iwama 1991). Since the mid-1980s, however, production of fish, crustaceans and shellfish by aquaculture has grown massively. Globally, aquaculture production has become the fastest-growing food production sector involving animal species. About 430 (97%) of the aquatic species presently in culture have been domesticated since the

Plate 3.1 In a photo taken from the International Space Station, sunglint reveals the density of aquaculture empoundments on the coast of Liaoning province, northeast China, in 2002 (Image Science and Analysis Laboratory, NASA-Johnson Space Centre, http://eol.jsc.nasa.gov/)

start of the 20th century (Duarte et al. 2007) and the number of aquatic species being domesticated is still rising rapidly. It was recently estimated that aquaculture provides 43% of all the fish consumed by humans today (FAO 2007a). Plate 3.1 is a satellite image showing the extent of development of aquaculture empoundments on the coast of Liaoning Province in northeast China.

The landings of fish from the world's oceans have gradually declined in recent years as stocks have been progressively overfished (Pauly et al. 2002). At the same time, demand for seafood has been steadily rising and, in parallel, aquaculture production has expanded significantly. Figure 3.1 depicts production from marine capture fisheries and aquaculture and illustrates the increasing production from aquaculture. This expansion of aquaculture is both a response to an increasing demand for seafood and, especially in the case of luxury products such as salmon and shrimp, an underlying cause of that rising demand. Table 3.1 shows the increase in world aquaculture production for the years 2000–2005.

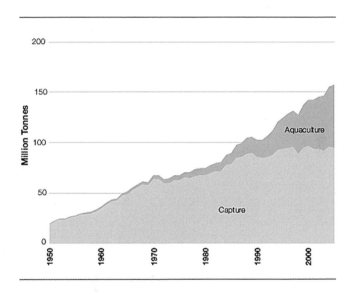

Fig. 3.1 Global fisheries and aquaculture production (all animals and plant species) 1950–2005 (FAO FISHSTAT database)

Table 3.1 World aquaculture production (excluding plants) for the years 2000–2005 (Adapted from FAO 2007a)

World production (million tonnes)	2000	2001	2002	2003	2004	2005
Marine aquaculture	14.3	15.4	16.5	17.3	18.3	18.9
Freshwater aquaculture	21.2	22.5	23.9	25.4	27.2	28.9

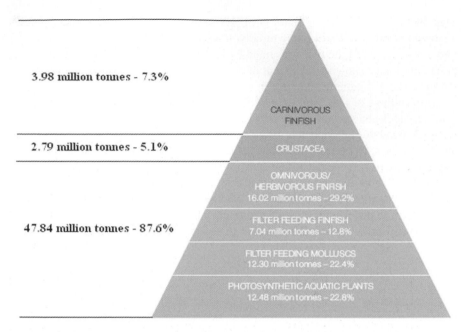

Fig. 3.2 Global aquaculture production pyramid (all animals and plants) by feeding habit and nutrient supply in 2003 (Adapted from Tacon et al. 2006)

The animal species that tend to dominate world aquaculture are those at the lower end of the food chain – shellfish, herbivorous fish (plant-eating) and omnivorous fish (eating both plants and animals) – with omnivorous scavenging crustaceans and carnivorous fish together contributing to less than 15% of total production (see Fig. 3.2). For example, carp and shellfish account for a significant share of the species cultivated for human consumption in developing countries (Naylor and Burke 2005). However, production of species higher in the food chain, such as shrimp, salmon and marine finfish, is now growing in response to a ready market for these species in developed countries (FAO 2007a; Naylor and Burke 2005). Methods of aquaculture production are described in Box 3.1.

Against a continuing background of diminishing and over-exploited marine resources, aquaculture has been widely held up as panacea to the problem of providing a growing world population with ever-increasing amounts of fish for consumption. With the expansion of the industry, however, the tendency has been for methods of production to intensify, particularly in the production of carnivorous species. This has resulted in many serious impacts on the environment and also in human rights abuses.

Box 3.1 Methods of aquaculture production

For freshwater aquaculture, ponds are either used or created and they are often located on areas of agricultural land. For the purposes of marine aquaculture, production takes place along the coast either in ponds, or in cages or net pens in the sea. Land-based systems include raceways (channels through which water from a natural source flows) or re-circulating systems in which fish are enclosed in tanks and through which treated water is re-circulated.

Different types of aquaculture are described as being extensive, semi-intensive or intensive. These descriptions refer to the input of food into the system:

- In extensive aquaculture, the farmed organisms largely take their nutritional requirements from the environment (Beveridge et al. 1997). However, nutrient-rich materials are often given to encourage the growth of algae on which the farmed species feed (Naylor et al. 2000). Traditional systems of aquaculture tend to be extensive and can be sustainable.
- In semi-intensive aquaculture, food from the environment is supplemented with fertiliser and/or food. This food is usually sourced from agricultural by-products, manures for example, or rice bran (Beveridge et al. 1997). Some fish protein in the form of fishmeal may also be used in semi-intensive aquaculture (Naylor et al. 2000).
- In intensive aquaculture, all or virtually all of the nutrition is provided directly from added feeds and/or fertiliser. Food is usually fishmeal (Beveridge et al. 1997). The farming of carnivorous species is generally intensive.

3.2 Negative Impacts of Aquaculture on People and on the Environment

The following two case studies discuss negative impacts associated with salmon and shrimp aquaculture. These case studies are far from exhaustive; rather, they provide examples that illustrate the wide spectrum of problems associated with the activities of aquaculture, and cast serious doubts on industry claims of sustainability. Note that a discussion of impacts of the aquaculture of tuna (i.e. tuna ranching) on wild tuna stocks in the Mediterranean is given in Chapter 6, Section 6.2.

3.2.1 Case Study 1: Shrimp Farming

Commercial shrimp farming has boomed. It began in the 1970s and grew rapidly during the 1980s. By 2001, 40% of shrimp sold were of farmed origin rather than wild caught (Goldburg and Naylor 2005).

3.2.1.1 Collection of Wild Juveniles as Stock for Aquaculture

The aquaculture of many species in the marine environment relies on juvenile fish being caught from the wild to supply stock rather than using hatcheries to rear them. Examples of aquaculture which collect juveniles from the wild include shrimp farming in south Asia and Latin America, milkfish in the Philippines and Indonesia, eels in Europe and Japan and tuna in South Australia and the Mediterranean (Naylor et al. 2000). In some cases the collection of wild juveniles has led to their over-exploitation. In addition, the practice may also result in the capture of juveniles of numerous other species, which are discarded and die.

Globally, it has been estimated that 65–75% of all shrimp juveniles (known as post larvae) used by shrimp farms are produced in hatcheries, but shrimp farms in many areas still rely on juveniles caught from the wild. According to Islam et al. (2004), natural stocks of shrimp are now over-exploited as a result of juvenile collection from the wild, and several reports suggest an extreme shortage of shrimp juveniles in some parts of the world. Furthermore, once caught, the shrimp juveniles only represent a small fraction of each catch – there is a large incidental catch (by-catch) and mortality of other species (Islam et al. 2004; Islam and Haque 2004). For example, the by-catch and mortality of numerous species has been reported in Honduras, India and Bangladesh:

- In Bangladesh, for each single tiger shrimp (Penaeus monodon) juvenile collected, there were 12–551 shrimp larvae of other species caught and wasted, together with 5–152 finfish larvae and 26–1,636 macrozooplanktonic animals.
- In Honduras, the reported annual collection of 3.3 billion shrimp juveniles resulted in the destruction of 15–20 billion fry of other species (Islam et al. 2004).
- In the Indian Sundarbans, each tiger shrimp juvenile only accounted for 0.25–0.27% of the total catch. The rest of the catch consisted of huge numbers of juvenile finfish and shellfish that were left aside on the beach flats to die (Sarkar and Bhattacharya 2003).

Islam et al. (2004) noted that the collection of shrimp fry not only posed serious impacts on regional biodiversity and aquatic community structure through such indiscriminate discard of juveniles but also by reducing the availability of food to other species in the food web, such as aquatic birds and reptiles.

3.2.1.2 Destruction of Habitat

Marine aquaculture for tropical shrimp and fish has typically used previously unexploited areas of land for pond construction (Beveridge et al. 1997). In many countries this has led to the irreversible destruction of thousands of hectares of

Fig. 3.3 Shrimp ponds and few remaining mangroves in bay of Guayaquil, Ecuador (Greenpeace/ Clive Shirley)

mangroves and coastal wetlands. Figure 3.3 shows a typical expanse of shrimp ponds in Guayaquil, Ecuador, with only a fraction of natural mangrove vegetation remaining.

Mangrove forests consist of trees and other plants that grow in brackish to saline tidal waters on mudflats, riverbanks and coastlines in tropical and subtropical regions. They are home to a diverse array of marine animals and some animals from inland; provide important nursery grounds for many marine and estuarine species such as finfish and shellfish, including commercially important species; stabilise coastlines from storm and tidal surges; and provide vital subsistence for coastal communities in many countries (for a discussion see Chapter 1, Section 1.5).

A review of aquaculture and mangrove destruction by Boyd (2002) suggested that human activities other than aquaculture have led to the majority of losses of mangrove forest. However, the literature clearly shows that coastal aquaculture, and in particular shrimp aquaculture, has itself caused substantial losses in mangrove habitat. For example:

- Beveridge et al. (1997) cited research published in 1991 that reported that, in the Philippines, 60% of the total reduction in mangrove areas was due to aquaculture. This was predominantly for shrimp aquaculture (Beardmore et al. 1997).
- In Bangladesh, it has been reported that more than 50% of the mangroves were lost, in particular for shrimp aquaculture (Das et al. 2004).

- In Vietnam, mangroves declined from 2,500 km² in 1943 to 500 km² in 1995, caused mainly by the encroachment of shrimp farms (Singkran and Sudara 2005).
- In Thailand, between 1961 and 1986, 38% of the total mangrove loss was attributed to aquaculture (Flaherty and Karnjanakesorn 1995). Another study in Thailand estimated that, between 1979 and 1993, 16–32% of the total mangrove area lost was converted to shrimp culture (Dierberg and Kiattisimkul 1996).
- In Ecuador, the Coordinator of Organizations for the Defense of Mangrove Forests (C-Condem) estimates that over 60% of its mangroves forests were lost in the second half of the last century. Between 1969 and 1992, Boyd (2002) estimated that 15–20% of the mangrove loss was caused by shrimp culture alone.

Destruction of mangrove habitat exposes large areas of soil to erosion and destroys former nursery grounds for aquatic organisms. Consequently, it leads to a reduction in species diversity and a decline in genetic diversity (Singkran and Sudara 2005; Beardmore et al. 1997). Islam and Haque (2004) noted that destruction of mangroves has caused a reduction in the natural production of fish and shrimp larvae. This reduction in juvenile shrimp, in turn, decreases the availability of shrimp juveniles for aquaculture farms and has resulted in the abandonment of farms. Furthermore, Flaherty and Karnjanakesorn (1995) highlight the potential for negative impacts to inshore fisheries due to the removal or modification of nursery grounds. The loss in wild fisheries stocks may be large. For example, in Thailand, it has been estimated that a total of 400 g of fish and shrimp are lost from fisheries for every 1 kg of shrimp farmed by aquaculture facilities developed in mangroves (Naylor et al. 2000).

The conversion of mangroves to shrimp farms can also lead to nutrients from the shrimp ponds draining into adjacent estuaries. This process can threaten estuarine animals, particularly fish (Singkran and Sudara 2005). In their natural state, mangroves act as sediment traps and stabilize water quality in the coastal zone. Their removal can be a threat to nearby coral reefs because corals require nutrient poor waters of low turbidity for their growth (Rönnbäck 1999).

Despite the widespread conversion of mangroves for aquaculture, these habitats are by no means ideal for aquaculture. This is because ponds reclaimed from mangrove become too acidic to support shrimp aquaculture within a few harvests. For instance, it has been estimated that the mean lifetime for a Thai pond is 7 years, although substantially shorter lifetimes are possible (Dierberg and Kiattisimkul 1996). As the decline in pond utility inevitably leads to abandonment, this may bring pressure to clear new areas and the whole 'boom and bust' cycle starts again (Naylor et al. 1998).

It has been noted that, with approximately 50% of the world's mangrove ecosystems already destroyed or transformed by human activity, the incremental cost of mangrove conversion to shrimp ponds is high (Naylor et al. 1998). Indeed, in order to protect coastal estuarine habitats and water quality for aquatic life, new shrimp farming in existing mangroves has been banned in Thailand. Even so, illegal use of mangroves for shrimp farms is still apparent and the topic has become very controversial (Singkran and Sudara 2005). In many Latin American countries, mangrove forests are protected by strict national environmental laws. Unfortunately, this has not impeded the shrimp farming industry, which has continued to occupy new mangrove areas illegally over the last 2 decades. This is still the case today. Boyd (2002) notes that most governments are coming to recognise the benefits of mangroves in their natural state and are beginning to regulate their use. However, there remains an urgent need to develop better policies and regulations regarding mangrove use and to enforce those regulations in a fair way.

3.2.1.3 Chemicals Used to Control Diseases

Intensive aquaculture greatly increases the risk of disease outbreaks among stock by concentrating many individuals in a small volume (high stocking density), maintaining continuous production cycles for many years and allowing wastes to accumulate in ponds or beneath cages (Pearson and Inglis 1993; Buchmann et al. 1995). As a consequence, a wide variety of chemicals and drugs may be added to aquaculture cages and ponds in order to control viral, bacterial, fungal or other pathogens (Gräslund and Bengtsson 2001; Wu 1995).

Pesticides and Disinfectants

Gräslund and Bengtsson noted in 2001 that there is generally a lack of information about the quantities of chemicals used in shrimp farming in Southeast Asian countries. However, based on knowledge of the types of chemical used there is a cause for concern. For instance, chemicals identified as being used at that time in Thai shrimp farms included copper compounds and triphenyltin, an organotin compound. These compounds are likely to leave persistent, toxic residues in sediments that can, in turn, cause negative impacts on the environment. In addition, copper is moderately to highly acutely toxic to aquatic life. The use of triphenyltin compounds had already been banned in some other Asian countries. A more recent survey of shrimp farms in Sinaloa, Mexico, reported that pesticides were not used (Lyle-Fritch et al. 2006).

Antibiotics

A range of antibiotics are in use worldwide in aquaculture to prevent or treat diseases caused by bacteria. With regard to the usage of antibiotics in aquaculture, the Food and Agricultural Organisation of the United Nations (FAO) has developed a Code of Conduct for Responsible Fisheries (FAO 1995). The Code indicates that preventative use of antibiotics in aquaculture should be avoided as far as possible and any use of antibiotics should preferably be under veterinary supervision (Holmström et al. 2003). Preventative (or prophylactic) use of antibiotics entails their use on a regular basis to prevent disease, rather than to treat disease when it occurs. Holmström et al. (2003) noted that, whereas for shrimp farming in general there is little published documentation on usage patterns of antibiotics, there was evidence that prophylactic use of antibiotics was a regular occurrence on many shrimp farms in Thailand. Such regular preventative application increases the risk of bacteria becoming resistant to the antibiotics in use, leading to serious problems if resistance is developed by a bacterial strain that can cause disease in the aquaculture stock.

Furthermore, there is a risk that bacteria which are pathogenic (cause disease) in humans could become resistant to an antibiotic that is used to treat the disease in humans. This could be a serious risk to public health (Miranda and Zemelman 2002).

Research has confirmed a number of instances in which the use of antibiotics in aquaculture has already led to the development of bacterial resistance. In Vietnam, one study found a relatively high incidence of bacterial resistance to antibiotics that were in use on shrimp farms, in samples of water and mud (Le et al. 2005). In the Philippines, bacteria from shrimp ponds were found to be resistant to four different antibiotics. Such multiple resistance was also reported to occur in a hatchery for shrimp aquaculture in India (Holmström et al. 2003). In Thailand, one of the factors which led to the collapse of the shrimp farming industry in 1988 was the indiscriminate use of antibiotics. This led to the development of resistant strains of bacteria which, in turn, were left free to cause disease in the shrimp (Holmström et al. 2003).

3.2.1.4   Depletion and Salinisation of Potable Water: Salinisation of Agricultural Land

Intensive shrimp farming in ponds requires the pond water to be brackish. Water must continuously be renewed and the salinity adjusted accordingly in the ponds. Up to 40% of the water in shrimp ponds is flushed out on a daily basis. This results in a high demand for seawater, freshwater, and brackish water resources. In some areas, this places an unsustainable demand on the freshwater supplies needed by communities for domestic use and food production (Public Citizen 2004). In addition, pumping fresh water from groundwater aquifers into shrimp ponds can result in a lowering of the water table. In turn, this causes seawater to seep in and water becomes unfit for consumption (Barraclough and Finger-Stich 1996).

Problems of salinisation and depletion of groundwater have been reported for many major shrimp-producing countries including Thailand, Taiwan, Ecuador, India, Sri Lanka, Indonesia and the Philippines (Environmental Justice Foundation 2004). In Sri Lanka, for example, it has been reported that 74% of coastal peoples in shrimp farming areas no longer have ready access to drinking water, due to excess salt in the water (Environmental Justice Foundation 2003).

Agricultural land can also become polluted by salinisation from seawater that has been pumped into shrimp ponds and is often flushed out within terrestrial environments (Barraclough and Finger-Stich 1996). The result can be increased soil salinity, which can prevent vegetable growth and kill plants used for cattle fodder (Environmental Justice Foundation 2003). For example, in Bangladesh there have been numerous reports of crop losses due to the salinisation of land following the onset of shrimp aquaculture (Environmental Justice Foundation 2004).

3.2.1.5 Human Rights Abuses

An Environmental Justice Foundation (EJF) report on shrimp farming in some less developed countries is a testimony to the human conflict and human rights abuses that have been suffered as a result of the setting up and running of this industry (Environmental Justice Foundation 2003). Although shrimp farming has been promoted by international financial institutions as a way of alleviating poverty, in reality this has often not been the case. Whilst a few entrepreneurs and investors have become rich, for many people shrimp farming has led to a degraded quality of life. Impacts associated with the industry include increased landlessness, decreased food security, child labour, intimidation, violence and murder.

Landlessness and Food Insecurity

The positioning of shrimp farms has often blocked coastal areas that were once common land to be used by many people. As a consequence, in areas of shrimp farming, access to fishing sites and mangrove forest resources for local people can become severely limited. There is often a lack of formalised land rights in such areas and this has led to large-scale displacement of communities from areas that have been inhabited for generations. The result has been landlessness and reduced food security for thousands of local families. In addition, farmers have also been displaced from their agricultural land because of the development of shrimp aquaculture Accounding to the Environmental Justice Foundation (2003), in some instances, displacement of people from land has been inflicted by invasion from gangs operated by shrimp farm owners or by cheap pay-offs from the state. A number of case studies of land seizures for shrimp farm construction are given in Box 3.2.

Intimidation, Violence, Rape and Murder

According to a report by the Environmental Justice Foundation (2003), non-violent protests about the shrimp farming industry have frequently been met with threats, intimidation and even violence from guards and musclemen associated with the shrimp industry, as well as false arrest and aggression from police. In at least

---

Box 3.2 Case studies of land seizures for shrimp farm construction

- Some Indonesian shrimp farms have been constructed following forced land seizures in which companies, supported by police and government agencies, provided either inappropriate compensation or none at all. Such cases have been reported from Sumatra, Maluku, Papua and Sulawesi.
- In Ecuador, reports indicate that there have been thousands of forced land seizures, only 2% of which have been resolved on a legal basis. Tens of thousands of hectares of ancestral land have allegedly been seized. This has often involved the use of physical force and the deployment of military personnel (Environmental Justice Foundation 2003).
- Between 1992 and 1998, many coastal people dwelling in the Gulf of Fonseca, Honduras, lost access to their traditional food sources and access to fishing sites because of encroachment on land by commercial shrimp farming companies (Marquez 2008).
- In Burma, the military has seized land without compensation in order to construct shrimp farms (Environmental Justice Foundation 2003).

---

11 countries, protestors have been murdered. In Bangladesh, it has been estimated that 150 people have been killed since 1980 in violent clashes related to shrimp farming. There are also cases of sexual harassment to women from guards at shrimp farms in Bangladesh, with 150 cases of rape reported in one district.

In some countries, the shrimp industry has become very powerful and is suspected to have tight links with individuals in governments, the police, military and judiciary. The perpetrators of violence in relation to the shrimp industry have rarely been brought to justice (Environmental Justice Foundation 2003).

3.2.2   Case Study 2: Salmon Farming

Farmed salmon are raised in hatcheries from eggs and are cultivated to market size in marine net pens. The industry has grown dramatically in recent years, with global production increasing fourfold between 1992 and 2002, such that it now exceeds the wild salmon catch by about 70% (Naylor et al. 2005).

3.2.2.1   Nutrient Pollution

Organic wastes from fish or crustacean farming include uneaten food, fecal matter, urine and dead fish (Goldburg et al. 2001). In the case of cage aquaculture (e.g. salmon farms), this waste matter enters marine waters directly. Waste from some pond aquaculture (e.g. shrimp farms) may also be deliberately released into the aquatic environment. Fish excreta and decaying food or fish release into the surrounding waters, among other things, sources of organic and inorganic nitrogen (including ammonia and nitrate) and phosphorous. These substances act, in turn, as nutrients and can support the growth of marine plants, including both macro-algae (seaweeds) and micro-algae (phytoplankton) (Scottish Executive Central Research Unit 2002). However, if discharged in excess, especially in poorly flushed areas, waters can become so enriched with nutrients that the result is nutrient pollution and the excessive growth of algae (termed eutrophication). The impacts of nutrient pollution, whatever the source of nutrients, can include (Goldburg et al. 2001; Scottish Executive Central Research Unit 2002):

- Foaming of seawater, and murky water
- Low dissolved oxygen levels
- Killing of wild fish or farmed fish or seabed animals
- Increased abundance of micro-algae possibly leading to harmful algal blooms
- Changes in marine food chains

Such effects of nutrient pollution have been reported to occur in the vicinity of salmon farming facilities. The quantity of nutrients discharged from aquaculture can be significant on a local scale. For example, according to literature cited by Naylor et al. (2003) a salmon farm of 200,000 fish releases an amount of nitrogen, phosphorous and fecal matter roughly equivalent to the nutrient waste in untreated sewage from 20,000, 25,000, and 65,000 people respectively. Many salmon farms in the Pacific Northwest and Norway contain four to five times that number of fish. Nutrient wastes from salmon farming has been a cause for concern among governments and some non-government organisations in Canada, Ireland, Norway and Scotland where wastes are released into what are considered to be otherwise unpolluted or sensitive coastal waters (Mente et al. 2006).

Effects on the Seabed

The most visible effects of nutrient pollution at salmon farms are those that impact on the seabed. When organic wastes reach the seafloor, oxygen can become depleted primarily through the activities of bacteria. Only a few animal species can survive these conditions and biodiversity in such areas therefore decreases. In severe cases, this can result in a 'dead zone', devoid of life beneath cages, and surrounded by an area of decreased animal diversity (Goldburg et al. 2001). Significant impacts have been reported to extend up to 100 m from cages, and more subtle effects up to 150 m away, although generally, the impact extends 20–50 m around the cages (Mente et al. 2006). For example:

- Research near finfish farms in the Bay of Fundy, Canada in the 1990s showed that diversity of animal fauna (macrofauna) was reduced close to farms throughout the area and, after 5 years of farms' operation, changes were documented up to 200 m away from cages (Fisheries and Oceans Canada 2003).

- In the west of Scotland, diversity of fauna was found to decrease around salmon farms (Mente et al. 2006).
- Research at eight salmon farms in Chile along a 300 km stretch of coastline showed that biodiversity was reduced by at least 50% on average in the vicinity of the farms. The loss of biodiversity seemed to be related to the quantity of organic matter and low oxygen levels in the sediments as well as the deposition of copper (Buschmann et al. 2006).

Even if severe impacts may be restricted to an area of a few hundred metres surrounding individual cages, the presence of multiple cages and/or farms in any particular area may contribute to greater cumulative impacts.

In an attempt to alleviate the problem of nutrient pollution, research is being conducted into cultivating seaweeds and shellfish near to farms because these species can use fish farm waste nutrients for growth (see Section 3.5).

Effects on Algae

Although aquaculture wastes release nitrogen and phosphorous into the water, they are not rich in silica. This creates conditions that are less favourable to diatoms and more favourable to the growth of other types of phytoplankton which are usually slow growing (dinoflagellates and cyanobacteria) (Mente et al. 2006). The rapid growth of such species as a result of nutrient pollution, in combination with other poorly understood factors, may lead to dense 'algal blooms' which can deplete oxygen at depth, reduce light penetration to other plants and, in some cases, even generate potent toxins. Such harmful algal blooms can thereby cause the death of marine plants and animals through a range of mechanisms. Some particularly harmful species are associated with shellfish poisoning in humans, which can occur when toxins produced by the algae are accumulated in shellfish such as mussels and oysters (Scottish Executive Central Research Unit 2002).

There is only limited research on the association between harmful algal blooms and salmon farming. In Chile, there have been increased reports of harmful algal blooms in the past 3 decades, and research on salmon farms indicated that the presence of farms has led to a significant increase in the abundance of dinoflagellates (Buschmann et al. 2006).

In the inter-tidal zone, nutrient pollution can result in an increase in green macroalgal (seaweed) mats that form a dense cover over the surface of the seabed. Most commonly this occurs with species of *Enteromorpha* and *Ulva*. An increase in *Enteromorpha* mats covering greater than 30% of the sediment has been found adjacent to salmon farms in the Bay of Fundy, Canada. This can have negative impacts on the growth rates of molluscs due to the creation of anoxic conditions within and below the mats (Fisheries and Oceans Canada 2003).

3.2.2.2   Escaped Farmed Salmon: Threats to Wild Fish

Individual populations of wild salmon are each specifically adapted to the rivers that they inhabit. This is reflected in a high genetic variability between different salmon populations. Naturally, there is also high genetic variability within each population. By contrast, farmed salmon have been selectively bred and have a low genetic variability (Naylor et al. 2005; Scottish Executive Central Research Unit 2002).

Unfortunately, farm-raised salmon have frequently escaped into the wild in vast numbers. Here they can compete with wild salmon for food and space, putting pressure on wild populations. Moreover, they can interbreed with wild fish. This is problematic because of their genetic differences. Their lower genetic variability can lead to the loss of unique gene pools in offspring, thereby potentially reducing their long-term adaptability to the environment. The offspring of wild salmon crossed with farmed salmon have been shown to be less fit than their parents (Naylor et al. 2005; Scottish Executive Central Research Unit 2002). One experiment cited by Naylor et al. (2005) showed that the lifetime success of wild fish crossed with farmed fish was significantly less than their wild cousins and that 70% of the embryos in the next generation died. The study demonstrates how interbreeding could drive vulnerable salmon populations to extinction. It is therefore of great concern that significant numbers of escaped farm salmon are surviving long enough to breed in the wild (Hindar and Diserud 2007). Continuing escapes may mean that the original genetic profile of the population will not re-assert itself (Goldburg et al. 2001).

What Is the Scale of the Problem?

Small-scale escapes of salmon from net pens arise routinely due to poorly main-tained pens or damage from seals. Moreover, net pens are open at the top such that, in stormy conditions, thousands of fish may escape. In just one incident in Norway in 2005, almost half a million fish escaped (Tidens Krav 2007). Naylor et al. (2005) cite literature which estimated that two million farm salmon escape each year into the North Atlantic.

Worldwide, over 90% of salmon which are farmed are Atlantic salmon (*Salmo salar*). In their native range, Atlantic salmon of farm origin are now successfully breeding in the wild, including in Norway (Hindar and Diserud 2007), Ireland, the UK and eastern North America. Outside of their native range in the Pacific, farmed Atlantic salmon have reportedly formed feral populations in rivers in British Columbia and in South America (Naylor et al. 2005). According to a study cited by Naylor et al. (2005), farmed salmon in Norway have been estimated to form 11–35% of the population of spawning salmon; for some populations they constitute greater than 80%.

What Impact Are Escaped Salmon Having?

Because farmed salmon are reproductively inferior to wild salmon, initially it was assumed that their chances of survival in the wild were poor. If they bred, natural selection should terminate their maladapted domestic traits. However, the sheer numbers of escaped fish, together with depleted wild salmon populations in the North Atlantic, means that natural populations may be dwarfed by the escapees such that inter-breeding could lead to reduced fitness in a population and increase the mortality of offspring (Naylor et al. 2005; Scottish Executive Central Research Unit 2002).

There is also the potential for direct competition for food and habitat. Farmed salmon juveniles are more aggressive than wild salmon and their behaviour can severely stress wild salmon, even increasing their mortality. The larger, more aggressive farmed fish can cause wild fish to move to poorer habitats, again increasing their mortality. In non-native regions, the farm escapees have competed for food and habitat with other fish in Pacific streams of North America and South America (Naylor et al. 2005).

What Can Be Done?

Naylor et al. (2005) notes that salmon farming companies have attempted to reduce the number of escapee fish by using stronger net materials, as well as using tauter nets to discourage seals. However, the number of escaping fish is still large and is having serious impacts on wild fish. One solution that has been suggested is to use land-based tanks or closed-wall sea pens so the fish are kept in closed containment. This would bring extra financial costs (Naylor et al. 2005) but, when put in the context of current threats to natural ecosystems, such costs are entirely justifiable.

3.2.2.3   Disease and Parasitic Infestations

There are concerns that disease from farmed species may be transferred to wild populations if farming is not contained from the wild environment. In salmon aquaculture, parasites and diseases are a major constraint on production (Naylor et al. 2003) and there is evidence that disease incidence in wild populations has been increased by salmon farming.

One example is sea lice (*Lepeophtherirus salmonis*), which are parasites that feed on salmon skin, mucus and blood. The lice can be a serious problem on farms and can even cause the death of fish (Goldburg et al. 2001). In the wild, sea lice generally have a low natural abundance and damage to salmon is minimal. Protection is afforded when salmon move from the sea to fresh water as most lice fall off in fresh water. However, when infestations occur on farms that are located in wild salmon habitat or on migration routes, wild salmon are at greater risk from infection (Naylor et al. 2003). Escaped farm salmon may also transmit the parasites directly to wild salmon. In British Colombia, there is evidence that pink salmon were affected by lice originating in farming areas (Naylor et al. 2003), while in Norway the highest infection levels in wild salmon have been found in salmon farming areas (Goldburg et al. 2001). In Chile, preliminary research also suggests that salmon farming can cause increases in sea lice infestations in native fish populations (see Buschmann et al. 2006).

In Canada, a study revealed that farm-origin lice caused 9–95% mortality in several wild juvenile pink and chum salmon populations (Krkošek et al. 2006). The study noted that the migratory cycles of salmon normally separate juveniles from adults and that this protects juveniles from contracting lice from the adults, important because juveniles are very susceptible to health impacts and death from lice infestation. Further work provided strong evidence that lice from farmed salmon have resulted in infestations in wild juvenile pink salmon that have depressed their populations (Krkošek et al. 2007). The authors suggested that, if the outbreaks continue, local extinction of pink salmon is certain. A 99% collapse in pink salmon abundance is expected to occur within their next four generations.

Sea lice can act as host in the transfer of a lethal disease called Infectious Salmon Anaemia (ISA) between fish. ISA has been found on salmon farms in Norway, Canada, Scotland, the United States and other countries. The disease was detected for the first time in 1999 in wild salmon in a Canadian river and in escaped farmed salmon in the same river. There were serious outbreaks of ISA on Chilean salmon farms in 2007, which necessitated a major culling operation (Fish Site News Desk 2007).

Another disease, furunculosis, is caused by bacteria. It spread to Norwegian farms from infected fish transported from Scotland in 1985. Escaped fish from infected farms caused the spread of the disease to wild salmon and, by 1992, it was detected in fish from 74 rivers (Naylor et al. 2005). Presently, this disease is no longer a problem in fish farming due to vaccination programmes (Scottish Executive Central Research Unit 2002).

3.2.2.4   Impacts on Marine Mammals and Birds

In Chile, sea lions (*Otaria flavescens*) have been found to attack farmed salmon net pens to feed. The expansion of salmon farming in Chile has caused increased mortality of sea lions due to their accidental entanglement in nets and by deliberate shooting by the farms. Deterrents include the use of acoustic devices to ward off the sea lions, but only the siting of anti-predator nets around the cages has resulted in a permanent reduction in attacks (FAO 2007c).

In Scotland, acoustic devices and anti-predator nets have been used to protect salmon net pens from seal attacks, though seals have also been shot. There is concern relating to the use of acoustic devices on cetaceans (dolphins, porpoises and whales) because these animals are highly sensitive to acoustic noise, whereas seals are less sensitive. For example, a Canadian study found that killer whales were excluded from a 10 km radius of an acoustic device. Therefore, while acoustic devices probably have no negative impact on seal populations, these devices may exclude cetaceans from a much larger area (Scottish Executive Central Research Unit 2002).

Birds attempting to prey on fish can become entangled in aquaculture nets (Australian Marine Conservation Society 2008) and may also be shot.

3.2.2.5 Human Rights Issues

In southern Chile, the salmon farming industry has grown rapidly since the late 1980s with high levels of foreign investment. It exports its product to western nations such as Japan and America (Phyne and Mansilla 2003; Barrett et al. 2002). In 2005, Chile produced nearly 40% of the world production of farmed salmon (Pizarro 2006), produced in farms such as that shown in Fig. 3.4.

In some countries, human rights abuses stem from the desire of aquaculture industry producers and processors to maximise profits within a highly competitive market, while meeting the low prices demanded by consumers. Presently, in the Chilean salmon farming industry, there are a number of serious human rights issues, as described below.

Fig. 3.4 Salmon farm near Puerto Chacabuco, Chile (Greenpeace/ Daniel Beltrá)

An Appalling Safety Record

One study has researched whether salmon farming in southern Chile has had negative or positive impacts on employees (Barrett et al. 2002). The study found that on salmon farms and in salmon processing plants, there were poor or non-existent health and safety regulations in place. For instance, on the salmon farms, working conditions were often cold, wet or unhygienic and there were no doctors or nursing staff. Another survey in 2004 found that there were a high number of accidents and job-related illnesses in the Chilean salmon industry, with 30% of workers suffering in that year (see Pizarro 2006). It has recently been reported that there have been more than 50 deaths in the Chilean salmon industry over the past 3 years, mostly of divers. In contrast, no-one has died in work-related incidents in the Norwegian salmon industry (Santiago Times 2007).

Low Wages and Long Working Hours

Barrett et al. (2002) and Pizarro (2006) reported that wages at Chilean salmon farms and processing plants were low. The average wage was insufficient for a single earner to raise a family of four out of poverty. The per capita income generated by the average wage is around the national poverty line (Pizarro 2006).

Barrett et al. (2002) reported that working hours in the processing plants could be long. For example, process workers at two sites in Chile worked an 8-hour day for 6 days a week and, during the high season, worked for 10–12 hours a day. In one of these plants, time missed because of illness had to be made up on Sundays.

Women Harassed

The number of women engaged in the salmon farming and processing plants is increasing. However, complaints of sexual harassment are constant, particularly at isolated farms. There is insufficient protection of maternity rights and an increasing number of related judicial cases. It has been reported that women who make use of their maternity rights later lose their jobs. It has been suggested that the reason for the high number of women in salmon farming and processing plants is due to the possibility of paying lower salaries (Pizarro 2006).

No Union Rights

Barrett et al. (2002) reported that, with the exception of one plant in southern Chile, there were no unions present to protect workers' rights in the salmon industry. This is because a strong union mentality does not generally exist in Chile due to the fact that, during the military dictatorship (between 1973 and 1989), union activity was particularly devastated and persecuted. The study noted that companies take advantage of this situation to attach a negative stigma to any type of union activity, and commented that it is the workers' fear of losing their jobs that prevents union pressure to fight for better wages. A 2007 news report on salmon farming in Chile noted that the labour organisation in Chile is fragmented and does not have the power adequately to protect workers' rights (Santiago Times 2007).

3.3 Use of Fishmeal, Fish Oil and Low Value or 'Trash' Fish in Aquaculture Feeds, and Associated Problems

Some types of aquaculture, notably the farming of carnivorous species such as salmon and shrimp, use fishmeal and fish oil in feeds. Farming of some species also relies on the use of whole fish of low market value (inappropriately termed 'trash fish'). Generally, fishmeal is used because it is digestible, energy-rich and a good source of protein, lipids (oils), minerals and vitamins (Miles and Chapman 2006), and is economically viable.

Fishmeal and fish oil are produced from the so-called 'industrial fisheries' that target small pelagic species. However, many of these industrial fisheries from which fishmeal and fish oil are produced are unsustainable. Furthermore, in some cases overfishing has resulted in negative ecological impacts (see Chapter 2, Section 2.5.3).

3.3.1 A Growing Demand

The quantity of fishmeal and fish oil utilised by aquaculture has increased over the years as the aquaculture industry has grown. For example, most recent estimates indicate that, in 2003, the aquaculture industry consumed 53.2% of the total world fishmeal production and 86.8% of world fish oil production (Naylor and Burke 2005; Tacon et al. 2006). The increasing trend for the use in fishmeal and fish oil for shrimp, salmonids and other marine finfish between 1992 and 2003 is shown in Table 3.2.

This increasing demand for fishmeal and fish oil by aquaculture has been met by diverting these products away from their use as feed for farmed animals, now

Table 3.2 Estimated use of fishmeal and fish oil in 1992 and in 2003 for three types of aquaculture products (Adapted from Tacon et al. 2006)

Aquaculture product	1992 Usage (t)		2003 Usage (t)	
	Fishmeal	Fish oil	Fishmeal	Fish oil
Salmonid	343,000	107,700	789,000	535,000
Shrimp	232,000	27,800	670,000	58,300
Marine finfish	180,000	36,000	590,000	110,600

increasingly restricted to starter and breeder diets for poultry and pigs. Fish oil was once used for hardening margarines and bakery products but is now mainly used in aquaculture (Shepherd et al. 2005).

If marine aquaculture production continues to rise, and the farming of carnivorous species is indeed set to increase, then the demand for fishmeal and fish oil could outstrip the current supply (Goldburg and Naylor 2005). However, some have the opinion that the use of fishmeal and fish oil by the aquaculture industry will decrease in the long term due to a number of factors, including prohibitively expensive prices (Tacon et al. 2006).

In recent years there has been much research and practical progress into substituting fishmeal with plant-based proteins, thereby lessening the inputs of fishmeal into diets, although the fraction of fishmeal, fish oil, and low-value fish used for the diets of carnivorous species remains high (see Section 3.4). Substitution with plant-based ingredients is positive, providing that this feed is derived from sustainable agriculture. However, the current shift to more plant-based feeds for aquaculture has not occurred fast enough to reverse the trend in fishmeal consumption, caused simply by an increase in the overall number of farmed carnivorous fish produced. For example, the quantity of wild fish required as feed to produce one unit of farmed salmon was reduced by 25% between 1997 and 2001, but the total production of farmed salmon grew by 60% over the same period (Naylor and Burke 2005). Given that many industrial fisheries are unsustainable, the aquaculture industry can never be seen to be sustainable unless it radically reduces its dependency on fishmeal and fish oil. Furthermore, the sustainability of using fishmeal, fish oil and low-value fish to feed carnivorous fish is unsustainable because there is an overall net loss of fish protein (Section 3.3.2) and the possibility of human food insecurity (Section 3.3.3).

3.3.2   Farming Carnivores: A Net Loss of Protein

It is often advocated by the industry that aquaculture will alleviate the pressure on stocks of wild fish in the oceans. This is not the case. Rather, the sustainability of farming some fish species is highly questionable because it results in a depletion rather than an increase in fish supplies, as a result of high feed inputs of fishmeal, fish oil or 'trash fish' in the diet. This is particularly the case for carnivorous species. For example, Naylor et al. (2000) reported that carnivorous species including salmon, other marine finfish and shrimp, require 2.5–5 times as much fish as feed (by weight) as the amount of fish produced. Thus, 1 kg of carnivorous fish produced can use up to 5 kg of wild fish in its production. For tuna that is caught wild and then fattened in ranches before harvesting, it has been reported that up to 20 kg of fish feed is required for each kilo of tuna produced (Volpe 2005).

Farming of carnivorous species that require such high inputs of wild fish as feed and produce a net loss of fish supplies cannot be viewed as sustainable. Only if the ratio of input of wild fish as feed to the output of cultured fish is less than one, is there an overall net gain in fish. To be classified as sustainable, not only should the conversion ratio of wild fish input to cultured fish output be less than one, but also the wild caught fish used as feed must come from fisheries that are sustainable.

3.3.3 Food Security Issues

The issue of diminishing rather than increasing net fish supplies is also one of food security since some species caught for fishmeal or classed by the industry as 'trash' fish can be important for human consumption (FAO 2007a). For example, in Southeast Asia and Africa, small pelagic (open water) fish such as those targeted by industrial fisheries are an important staple in the human diet (Sugiyama et al. 2004). Demand for such fish is likely to grow as human populations increase, bringing them under further pressure from both aquaculture and direct consumption (Naylor et al. 2000).

The increased demand for use in aquaculture of high value carnivorous species and/or for livestock feeding has led to increases in prices of low-value fish ('trash' fish) and this may mean that the rural poor can no longer afford to buy it (Tacon et al. 2006). Without intervention to prevent this from happening, economics rather than human need will drive the market supply. With these factors in mind, the FAO has recommended that there is a "need for governments within major aquaculture-producing countries to prohibit the use of 'trash fish' or low value fish species as feed for the culture of high value fish or shellfish species, and in particular within those countries where 'trash fish' is consumed directly by the rural poor" (Tacon et al. 2006).

3.4 Moving Towards More Sustainable Aquaculture Feedstuffs

As the aquaculture industry has grown, there been a concurrent rapid expansion in aquafeed production (Gatlin et al. 2007). The growth and intensification of aquaculture in some countries, together with the increased farming of carnivorous species, has caused a rise in demand for fishmeal and fish oil for such aquafeeds. Further increases in the use of finite fishmeal and fish oil resources for aquaculture could, however, simply be impossible. It is already apparent that the industrial fishing of many stocks is unsustainable (see Chapter 3, Section 3.5.3.1), and the anticipated growth of aquaculture could outstrip supplies of fish for aquafeeds within the next decade. Consequently, it has been recognised for many years by the aquafeed industry that the use of more plant-based feedstuffs, rather than fishmeal and fish oil, is essential in the future development of aquaculture (Gatlin et al. 2007).

The price of fishmeal has also been an important driving force. It has been noted that as the price of fishmeal increases, there is a considerable incentive for the aquaculture industry to innovate by, for example, substituting with plant-based ingredients (Kristofersson and Anderson 2006). In recent years, there has also been concern about elevated levels of persistent environmental contaminants present in fish oils, especially chlorinated dioxins (PCDD/Fs) and polychlorinated biphenyls (PCBs). This has increased pressure on feed manufacturers to produce oils with lower levels of these chemicals and thereby created an even greater interest in the use of vegetable oils (Scottish Executive Central Research Unit 2002).

Research on the reduction of fishmeal and fish oil in aquafeeds has focused on identifying and using products that can keep up with aquaculture growth. This has included using plant-based ingredients (Section 3.4.1), and fish trimmings from the processing of fish for human consumption (Section 3.4.3). Some aquaculture systems allow for the build up of a suspended 'floc' material (known as microbial floc or bio-floc), which is typically composed of phytoplankton, zooplankton, bacteria, protozoans, micro-algae and detritus (Conquest and Tacon 2006; Serfling 2006; Verdegem et al. 2006). Micobial flocs provide an additional feed source for the species being farmed and have a further advantage of mediating water quality by reducing levels of ammonia and nitrate in the water (Avnimelech 2006; Wasielesky et al. 2006).

Plant-based products are, to some extent, already used widely in aquaculture and research is ongoing to investigate their suitability in the diets of individual fish species. The plant products utilised in aquaculture are protein-rich oilseed and grain by-product meals, and include soybean, rapeseed, corn gluten, wheat gluten, pea and lupin meals, palm oil, soybean oil, maize oil, rapeseed oil, canola oil, coconut oil, sunflower oil, linseed oil and olive oil (see Tacon et al. 2006).

3.4.1 Utilisation of Plant-Based Products

It is important to note that if the use of plant-based feeds in aquaculture is to be sustainable, they must be sourced from agriculture that is sustainable. Among other requirements, sustainable agriculture precludes the use of any genetically modified (GM) crops. The use of GM plants creates its own dangers in terms of food and environmental safety. The process of inserting novel genes into plants or other organisms can cause unintended deletions or re-arrangements of existing genes or change the regulatory function of genes, with unpredictable results; for example, it is possible that new toxins or allergens may be produced. GM crops currently being grown in various parts of the world, including soya, corn and canola, have already caused environmental damage and contamination of conventional and organic crops (Greenpeace and Gene Watch UK 2007). There also remains many unresolved food safety concerns (Greenpeace 2007c). Thus, GM plants (or indeed GM fish, which have also been proposed) present additional environmental and health concerns, not solutions.

To be suitable for use in aquaculture feeds, plant feedstuffs must fulfill criteria of being widely available and cost-effective to produce, and must provide an adequately nutritious diet so as to produce high-quality fish flesh that will deliver human health benefits (Gatlin et al. 2007). A recent study has identified plant-based feedstuffs that are already used and/or show particular promise for the future (Gatlin et al. 2007). These include soybean, barley, canola, corn, cottonseed and pea/lupins.

It must be stressed, however, that plant products can have nutrient profiles that are not entirely suitable for fish and may contain bioactive compounds that are also not favourable. These are commonly referred to as anti-nutritional factors and can preclude the use of plant feedstuffs in diets at high concentrations. Gatlin et al. (2007) discussed processing methods which can help in this regard, as well as the possibility of using supplementation where nutrients are lacking. For example, nearly all plants contain phytic acid, a compound that is not digestible by fish. A recent study reported that, to counteract this problem, the enzyme phytase can be supplemented in feeds when they are formulated. This improves the utilisation of plant-based protein by fish, thereby positively affecting their growth (Gabriel et al. 2007). It has also been suggested that selective breeding of fish can be used to improve the ability of fish to use plant proteins (e.g. Quinton et al. 2007).

3.4.1.1 Feeding Herbivorous and Omnivorous Species

Tacon et al. (2006) reported that the best results to date for utilising plant feed in aquaculture feed is for herbivorous or omnivorous fish (carps, tilapia, milkfish, channel catfish). Total dietary fishmeal replacement has been possible with these species, without negative impacts on growth or feed efficiency. Rearing such species in this way suggests a more sustainable future for aquaculture provided that the feeds themselves are produced through sustainable agriculture.

3.4.1.2 Feeding Carnivorous Species

For carnivorous fish species, the proportion of fishmeal and fish oil in diets can be reduced by at least 50%, but complete substitution with plant-based ingredients has not been possible for commercial production. The level of fishmeal in diets for salmon is now commonly about 35% while the level of fish oil is about 25% (although these proportions vary somewhat between different countries).

The basic problems encountered in trying to replace all fishmeal and fish oil for carnivorous species are not only limited to concerns regarding anti-nutritional factors, but also include the lack of essential amino acids such as lysine and methionine and of the essential fatty acids eicosapentaenoic acid (EPA) and docosahexaenoic acid (DHA) (Tacon et al. 2006; Scottish Executive Central Research Unit 2002). The amino acids that are lacking can be added to the diet, but EPA and DHA fatty acids are more problematic. Fish is considered to be an important source of DHA and EPA (omega 3) fatty acids in the human diet, but these fatty acids are significantly reduced in fish that have been fed with plant oils instead of fish oil.

Nevertheless, recent research has shown that by using plant oil-based diets during the fish growth phase and switching to a fish oil-based diet during the period prior to slaughter, the fatty acid composition that is beneficial to human health is restored in the fish flesh. The use of such finishing diets has been suggested as a suitable way to deliver the required fatty acid content in farmed fish (Pickova and Mørkøre 2007). However, even though fish oil use could be reduced by this method, it seems unlikely that it can be replaced completely.

Recent research suggests that marine shrimp can be fed largely on plant-based feeds. Amaya et al. (2007) reported that Pacific white shrimp (*Litopenaeus vannamei*) could be fed a diet consisting of soy and corn ingredients instead of fishmeal without adverse impacts on shrimp growth. The plant-based diet did, however, contain 1% squid meal and fish oil. The study suggested that further research is needed to evaluate the replacement of fish oil and to evaluate potentially limiting nutrients in such diets. Another study also reported that growth of the Pacific white shrimp fed on an entirely plant-based diet (with no fishmeal or fish oil) was no different from shrimp fed on a fishmeal and fish oil diet (Browdy et al. 2006). However, the plant-fed shrimp had lower levels of the same two key fatty acids EPA and DHA. The authors of the study suggested that it would be possible to add supplements to remedy the problem, although it is not known whether this modification to the feed would be cost-effective. It was also suggested that further research could be conducted into achieving the desired fatty acid content of the shrimp by using finishing diets that contain fishmeal/fish oil and are given for a period shortly before harvesting.

3.4.2   Utilisation of Fish Trimmings

When fish for human consumption is filleted and processed for the market, more than half the fish is considered waste. Such fish trimmings can be used in the production of fishmeal by the aquaculture industry. In 2002, it was estimated that about 33% of the raw material supplied to the fishmeal and oil sector in Europe came from fish trimmings (Huntington 2004b). It has been estimated that the use of fish trimmings, or processing scraps from sustainable fisheries, could produce marine protein and oil yielding up to 20% of the world supply (Hardy 2007).

In some cases, the organic aquaculture sector is utilising fish trimmings as feed. For example, certification of organic Scottish salmon by the Soil Association specifies that all of the fishmeal, and the majority of the oil, comes from trimmings of fish caught for human consumption (Raven 2006).

The use of fish trimmings from fish caught for human consumption can be seen as more sustainable than using normal fishmeal, in that a waste product is being used. However, unless the fishery from which the fish trimmings come is itself sustainable, the use of fish trimmings cannot be seen as sustainable because it perpetuates the cycle of the over-exploitation of fisheries.

3.5   Moving Towards More Sustainable Aquaculture Systems

In order for aquaculture operations to move towards sustainable production, the industry needs to recognise and address the full spectrum of environmental and societal impacts caused by its operations. Essentially, this means that it will no longer be acceptable for the industry to place burdens of production (such as the disposal of waste) onto the wider environment.

In turn, this implies moving towards closed production systems. For example, in order to prevent nutrient pollution, ways can be found to use the nutrients present in waste products beneficially. Examples include integrated multi-trophic aquaculture (IMTA, including aquaponics) and integrated rice-fish culture.

In the IMTA system, the waste products and nutrients of fed species (finfish or shrimp) are utilised as food by other species that function at a different level of the food chain (trophic level). Economically important species that fall into this category include plants, such as seaweed, and shellfish. In an IMTA system, seaweeds extract the dissolved inorganic nutrients while shellfish extract particulate organic matter (Chopin 2006). In essence, IMTA systems aim to balance waste production and extraction, and thereby mimic natural ecosystem functions as much as possible (Neori et al. 2004, 2007).

Modern IMTA systems have been developed using ideas from traditional aquatic polyculture, defined as the culture of more than one species together. The difference between the systems is that IMTA requires the cultivation of species from different levels of the food chain, thereby reducing waste products, whereas polyculture can involve the co-cultivation of any species. Some aquatic polyculture has been practiced in China for millennia, such as the co-cultivation of rice and fish (Neori et al. 2004). Today, some Asian marine polyculture in coastal waters can be classified as IMTA since it uses wastes from fish cages to enhance the growth of adjacent cultures of shellfish and seaweeds (see Neori et al. 2007).

Species involved in IMTA systems include fish or shrimp integrated with vegetables, microalgae, shellfish and/or seaweeds (Neori et al. 2004). IMTA can be set up in coastal waters, in ponds or in land-based systems and can be highly intensified (Chopin 2006; Neori et al. 2004). Land-based systems which use the waste products of fish/shrimp culture as fertiliser for growing vegetables, known as aquaponics, is a variation of the IMTA concept.

It has been suggested that seaweed-based IMTA systems offer a more sustainable way forward for mariculture (marine aquaculture). Seaweeds filter waste nutrients from fish/shrimp culture (particularly carbon, nitrogen and phosphorous) and add oxygen to the seawater, thereby restoring water quality. Seaweeds can be cultured for food or other uses and can also act as a nutrient source for other co-cultured species such as abalone and sea urchins. The growth of seaweed on mariculture effluents has been reported to be superior to that on fertiliser-enriched clean seawater. Because ecological harm can be caused by the introduction of non-native species it is important that the seaweed used in IMTA systems should be a native species. Ideally the seaweed would be a species that would be of ecological value, in terms of removing excess waste products, as well as of economic value (Neori et al. 2004).

The use of IMTA systems is likely to become a way of negating costs for the 'polluter pays' charges. For example, Denmark is reconsidering more finfish aquaculture development only on the condition that there is adequate planning for bioremediation and use of bio-filters (seaweed and shellfish). In other words, the use of extractive species is now a necessity for the licence to operate in Denmark (Chopin 2006).

3.5.1 Examples of IMTA Systems

• SeaOr Marine Enterprises on the Israeli Mediterranean coast is a modern, intensive, land-based mariculture farm that cultivates marine fish (gilthead seabream), seaweed (*Ulva* and *Gracilaria* spp.) and Japanese abalone. Effluent waste from the fish culture is utilised for growth by the seaweed. In turn, the seaweed is fed to the abalone (Neori et al. 2004).

• Aquaponics involves using the effluent from fish farming as a nutrient source for the growing of vegetables, herbs and/or flowers. This negates the cost of a bio-filter used for other re-circulating aquaculture systems, and is more environmentally sustainable. The development of aquaponic technology since the 1980s has resulted in viable systems of food production. Plants such as lettuce, herbs, watercress, spinach, tomatoes and peppers are produced hydroponically (without soil, in a water medium) in greenhouses. In North America, the most common form of aquaponics farms freshwater tilapia (Diver 2006).

When the fish being farmed in IMTA systems are carnivorous and require feeding with fishmeal, fish oil or 'trash' fish, the sustainability of this aquaculture is called into question. Common sense dictates that it is important that there is a shift towards the cultivation of omnivorous or herbivorous species which do not require fish-based feeds and that these are co-cultured in IMTA systems in which effluent waste discharges are controlled, and utilised beneficially (by, for instance, seaweeds, vegetables and shellfish). It is therefore clear that, in order to expand sustainably, the industry needs to expand research and development on herbivorous and omnivorous fish (such as carp, milkfish, gray mullet, and catfish). Ideally, sustainable IMTA aquaculture would aim to develop closed systems, as open water systems may still carry a risk of nutrient pollution.

3.5.2 Integrated Rice-Fish Culture

A promising form of (land-based) aquaculture is the production of fish in rice fields, known as integrated rice-fish culture. This system optimises the use of land and water and is benefited by synergies between fish and plant. Rice-fish culture in China dates back to AD 220 and today is also practised in Egypt, Indonesia, Thailand, Vietnam, Bangladesh and Malaysia, among other countries. However, the extent of its use is presently rather marginal. Constraints to its wider use include a lack of education of farmers in the necessary skills, and it has been suggested that policy-makers need to provide more active support in this area (Frei and Becker 2005). It is important to note that integrated rice–fish culture is crucial for local food security, rather than representing a method for supplying exports to supermarkets in developed countries.

3.6   Recommendations

Any aquaculture that takes place needs to be sustainable and fair. For aquaculture systems to be sustainable, they must not lead to natural systems being subject to degradation caused by:

1. An increase in concentrations of naturally occurring substances
2. An increase in concentrations of substances, produced by society, such as persistent chemicals and carbon dioxide
3. Physical disturbance

In addition, people should not be subject to conditions that systematically undermine their capacity to meet their basic needs for food, water and shelter.

In practical terms, these four conditions can be translated into the following recommendations according to Greenpeace (see Allsopp et al. 2008).

Use of Fishmeal, Fish Oil and Low Value or 'Trash' Fish: To reduce the pressure on stocks caught for fishmeal and fish oil, there needs to be a continued move towards sustainably produced plant-based feeds. Cultivating fish that are lower down the food chain (herbivores and omnivores rather than top predators) and that can be fed on plant-based diets is key to achieving sustainable aquaculture practices. Industry must expand its research and development on herbivorous and omnivorous fish that have strong market potential and suitability for farming.

In more general terms, there is an urgent need for fisheries management to shift towards an ecosystem-based approach wherein a global network of fully protected marine reserves covering at least 40% of the oceans is established, together with sustainable fisheries management outside of the reserves (Roberts et al. 2006). This is key to achieving sustainable fisheries.

Greenpeace considers the culture of species that require fishmeal or fish oil-based feeds derived from unsustainable fisheries and/or which yield conversion ratios of greater than one (i.e. represent a net loss in fish protein yield) as unsustainable. Plant-based feeds should originate from sustainable agriculture, and sources of omega 3 should be algal derivatives, grape seed oils, etc.

Nutrient Pollution and Chemical Pollution: To reduce nutrient wastes, there is great potential for the development of integrated multi-trophic aquaculture (IMTA) systems, aquaponics and integrated rice-fish culture.

Greenpeace considers aquaculture that results in negative environmental impacts in terms of discharges/effluents to the surrounding environment as unsustainable.

Escapes of Farmed Fish to the Wild: To overcome these problems it has been suggested that enclosed bag nets or closed wall sea-pens should be used to prevent fish from escaping, or that land-based tanks should be used (Naylor and Burke 2005). Ultimately, land-based tanks are the only option if the goal is to eliminate any risk of escapes which might otherwise occur as a result of hurricanes or other extreme weather events at sea. It is crucial to use native rather than exotic species (Pérez et al. 2003).

Greenpeace recommends that only species that are native should be cultivated in open water systems, and then only in bag nets, closed wall sea-pens or equivalent closed systems. The cultivation of non-native species should be restricted to land-based tanks.

Protection of Local Habitat: Some aquaculture practices have had serious negative impacts on local habitat. Aquaculture practices must be set up in a way that provides for the protection of coastal ecosystems and local habitats. In addition, no new aquaculture practices should be permitted in areas that are to be designated as marine reserves, and any existing aquaculture operations within such areas should be phased out.

Greenpeace considers aquaculture that causes negative effects to local wildlife (plants as well as animals) or represents a risk to local wild populations as unsustainable.

Use of Wild Juveniles: The use of wild-caught juveniles to supply aquaculture practices, particularly some shrimp aquaculture, is destructive to marine ecosystems.

Greenpeace considers aquaculture that relies on wild-caught juveniles as unsustainable.

Transgenic Fish: The physical containment of genetically engineered fish cannot be guaranteed under commercial conditions and any escapes into the environment could have devastating effects on wild fish populations and biodiversity (Anderson 2004).

Greenpeace demands that genetic engineering of fish for commercial purposes should be prohibited.

Diseases: *Greenpeace recommends cultivation at stocking densities that minimise the risk of disease outbreaks and transmission and, therefore, minimise requirements for therapeutic treatments.*

Resources: *Greenpeace considers aquaculture that depletes local resources, for example, drinking water supplies and mangrove forests, as unsustainable.*

Human Health: *Greenpeace considers aquaculture that threatens human health as unfair and unsustainable.*

Human Rights: *Greenpeace considers aquaculture that does not support the long-term economic and social well-being of local communities as unfair and unsustainable.*

Chapter 4
Pollution

Abstract Many human activities cause the release of synthetic hazardous substances into the marine environment or the redistribution of naturally occurring substances. This can result in the contamination of the marine environment with substances that have a deleterious effect on marine life. This chapter selects and discusses some pollutants that are currently of concern in the marine environment.

Keywords Chemicals, brominated, PBDE, TBBPA, fluorinated, PFCs, persistent organic pollutants, POPs, toxicity, radioactivity, eutrophication, oil, dead zones, bioaccumulation, plastic debris, alien species, regulation.

Chemical Pollution: Two groups of chemicals are discussed here – brominated flame retardants and perfluorinated compounds – to illustrate the ongoing problem of contamination of the oceans with persistent organic pollutants (POPs) more generally. Studies show that certain brominated flame retardants which are now in widespread use contaminate the body tissues of many marine animals. For instance, these chemicals have been found in fish from many regions, in seabirds from the Arctic, in harbour porpoises and seals, in sperm whales from the deep oceans and in polar bears. Similarly, pollution with perfluorinated chemicals is also widespread, with contamination being present in the body tissues of many marine species. While certain brominated flame retardants and perfluorinated chemicals are under consideration by the United Nations Stockholm Convention on Persistent Organic Pollutants, as yet, no action has been taken towards a global phase out.

Radioactive Pollution: Contamination of the environment by artificial radio-nuclides is of concern because they can act as potent carcinogens and mutagens. Presently, the most prominent point sources of radioactive pollution to the oceans are from the Sellafield (UK) and La Hague (France) nuclear reprocessing plants.

Nutrient Pollution and Marine Dead Zones: Marine areas devoid of life have been termed 'dead zones'. There are now about 200 dead zones in the oceans, some of which have been linked to excessive nutrient input resulting from human activities.

Oil Pollution: Large oil spills to the marine environment can be catastrophic for wildlife, and health impacts can be long-lasting if the oil pollution persists.

M. Allsopp et al., *State of the World's Oceans,*
© Springer Science+Business Media B.V. 2009

This section discusses the initial and long-term impacts of two major oil spills, the Exxon Valdez in 1989 in the Gulf of Alaska and the Prestige in 2002 off the coast of northern Spain. In both cases thousands of birds were killed initially and, in the longer term, some bird and other animal populations continue to suffer from impacts on their health.

Plastic Debris: Marine debris, especially plastics and other man-made materials, has become a pervasive pollution problem affecting all of the world's oceans, from polar regions to the equator.

Serious injuries and deaths resulting from marine debris have been recorded for numerous marine animals and birds, either because they become entangled in it or they mistake it for prey and eat it. Entanglement in old fishing gear, six-pack rings and fishing bait box strapping can cause death by drowning, suffocation, starvation, and as a result of physical injuries. When ingested, waste such as plastic bags, pellets and fragments of larger items can block the digestive tract or fill the stomach, resulting in malnutrition, or starvation and death.

Pollution in the Marine Environment: This chapter discusses some of the pressing issues of marine pollution. The discussion of these topics is by no means exhaustive – that would be beyond the scope of this chapter. Instead, the commentary on each type of pollution gives specific examples, which are illustrative of the overall nature of the problem.

4.1   Chemical Pollution

One group of synthetic chemicals that have become widespread environmental pollutants and are of great concern are persistent organic pollutants, or POPs. These chemicals have the properties of being persistent (long-lived) in the environment, bioaccumulative (build up in the tissues of animals) and are toxic. They also undergo transport over large distances from their point of origin and consequently become widespread in the environment.

Many POPs have been listed by a United Nations (UN) global treaty designed to protect humans and the environment – the Stockholm Convention on Persistent Organic Pollutants (POPs). The Convention entered into force in May 2004. Implementation of the Convention requires governments to take measures to eliminate or reduce the release of POPs into the environment. Chemicals included by the Convention include the well-known dioxins and PCBs as well as several organochlorine pesticides (Stockholm Convention on Persistent Organic Pollutants 2008).

Presently, there are a number of other chemicals that are proposed for the POPs list of the Stockholm Convention or are under consideration for inclusion. These chemicals include certain brominated flame retardants and perfluorinated chemicals. In the following two sections, contamination of the marine environment by certain brominated flame retardants and perfluorinated chemicals is discussed to illustrate the problem of ongoing POPs pollution in the oceans. Note that contamination of the marine and terrestrial environment by other POPs which are currently listed by the Stockholm Convention, such as PCBs and dioxins, are widely discussed in the scientific literature, have been the subject of several other Greenpeace reports (e.g. Allsopp et al. 1999) and are not discussed here. Similarly, contamination of the marine environment with heavy metals is also not covered by this review.

4.1.1   Brominated Flame Retardants

Flame retardants are substances which are added or applied to a material to increase the fire resistance of that product (Alaee et al. 2003). One group of flame retardants, known as the brominated flame retardants, are a chemically diverse group of brominated organic compounds (Vorkamp et al. 2004a).

There are more than 75 different brominated flame retardants recognised commercially, but five comprise the vast majority of current production (Birnbaum and Staskal 2004). These are three commercial formulations of polybrominated diphenyl ethers (PBDEs), and hexabromocyclododecane (HBCD) and tetrabromo-bisphenol A (TBBPA).

With regard to all PBDEs, there are 209 potential different chemicals that are classified into ten groups according to their average bromine content (i.e. mono-BDE, di-BDE, tri-BDE etc. up to deca-BDE) (Alaee et al. 2003). The three commercial PBDE formulations mentioned above contain mainly either penta-BDE, octa-BDE or deca-BDE and consequently they are commonly referred to as PentaBDE, OctaBDE and DecaBDE. The EU recently banned the use of PentaBDE and OctaBDE formulations (EC 2003b) and China banned PBDEs in electrical and electronic goods, but there is not a global ban in place on these chemicals. In addition, DecaBDE, HBCD and TBBPA remain in widespread use.

4.1.1.1   Uses of Brominated Flame Retardants

PBDEs are widely used in plastics, textiles, paints and electronic appliances including computers, televisions and other electric household equipment (Ueno et al. 2004). HBCD is used in polystyrene foams, which are utilised as thermal insulation in buildings, in upholstery textiles, and to a minor extent in electrical equipment housings (Covaci et al. 2006).

About 90% of TBBPA is used in the production of epoxy and polycarbonate resins and the main application of epoxy resins is in the manufacturing of electronic printed circuit boards. About 10% of TBBPA is used in acrylonitrile-butadiene-styrene (ABS) resin and polystyrene.

When incorporated into products, both PBDEs and HBCD are used as additives and are simply blended with polymers rather than chemically bound. This means that they may separate or leach from the surface of products into the environment. On the other hand, TBBPA is more commonly used as a reactive component and is bound to materials chemically although there are some additive uses for TBBPA as well (Alaee et al. 2003; de Wit 2002). However, even products containing TBBPA in reactive form have been shown to release TBBPA into the environment (Birnbaum and Staskal 2004).

The production of brominated flame retardants has increased dramatically over the past 20 years (Birnbaum and Staskal 2004). For instance, global usage in 1990 was 145,000 t and grew to 310,000 t by 2000 (Alaee et al. 2003). Birnbaum and Staskal (2004) cited literature from 2001 which reported that Asia consumed an estimated 56% of the total market demand of brominated flame retardants, the Americas 29% and Europe 15%. The use of DecaBDE and TBBPA account for about 50% of the world's consumption of brominated flame retardants (Brigden et al. 2005; Morris et al. 2004).

4.1.1.2   Sources of Environmental Contamination

There are a number of different pathways by which brominated flame retardants may enter the environment, including emissions during production or use, leaching from finished products, or following disposal (Covaci et al. 2006; de Wit 2002). Sewage sludge has been found to contain PDBEs, HBCD and TBBPA indicating sources coming from users, households and industries generally. The application of sewage sludge on agricultural land in many countries further increases the likelihood of their remobilisation in the environment. Point sources from industries using brominated flame retardants cause local hotspots of pollution (Law et al. 2006a). Waste electronic equipment such as computers can contain brominated flame retardants and much ends up in Africa and Asia where it is dumped or 'recycled' by a largely unregulated industry (Cobbing 2008). Not surprisingly, the local environment has been found to be contaminated with brominated flame retardants in China and India where such processing of electronic waste occurs (Brigden et al. 2005; Leung et al. 2007).

The presence of PBDEs, HBCD and TBBPA in samples from living organisms in the Arctic and the presence of PBDEs in air samples from Europe, USA and the Arctic, suggests that these compounds undergo long-range transport in air (Covaci et al. 2006; de Wit et al. 2006; de Wit 2002). Research on levels of PBDEs and HBCD in Arctic polar bears suggests that western Europe and eastern North America act as sources of these chemicals to Arctic regions (de Wit et al. 2006).

4.1.1.3   Bioaccumulation

Studies have shown that PBDEs bioaccumulate in the tissues of living organisms. The lower brominated PBDEs (four to seven bromines) are the most bioaccumulative and persistent (Birnbaum and Staskal 2004). HBCD is also bioaccumulative. Studies show that PBDEs and HBCD are contaminants in the body tissues of many species of marine animals. With regard to TBBPA, a study on laboratory rats suggested that it has the potential to bioaccumulate but a study on fish and oysters suggested that it did not. It has however been found in the tissues of some wildlife species, although research is limited (Law et al. 2006b; Morris et al. 2004; de Wit et al. 2006).

Brominated flame retardants may be passed from mother to developing young. For example, Vorkamp et al. (2004a) cited research on pilot whales (*Globicephala melaena*) which found higher levels of PBDEs in juveniles than in adults. This suggested lactational transfer of the chemicals from mother to young. In humans, PBDEs and HBCD have been detected in breast milk (Birnbaum and Staskal 2004). In harbour seals (*Phoca vitulina Richardsi*), the transfer of PBDEs across the placenta to the developing foetus was suggested after tissue from a dead pup found in the birth canal of a dead seal was found to contain the chemicals (She et al. 2002).

There is evidence that some PBDEs increase in concentration (biomagnify) through marine food chains. This means that the highest tissue levels are found in predatory animals at the top of the food chain. For example, a study in Greenland showed that ringed seals (*Phoca hispida*) had higher tissue levels than the shorthorn sculpin (*Myoxocephalus scorpius*), a fish, and that polar bears (*Ursus maritimus*) had higher tissue levels than the seals (Vorkamp et al. 2004a). Burreau et al. (2000) reported biomagnification of some PDBEs between certain fish species of the Baltic Sea and northern Atlantic. A study in coastal Florida reported concentrations of PBDEs in dolphins and sharks were —one to two orders of magnitude higher than in fish from lower down the food chain, which indicated biomagnification (Johnson-Restrepo et al. 2005).

Research on levels of HBCD in wildlife also shows that levels are often elevated in species at the top of food chains, a phenomenon clearly suggesting biomagnification (Covaci et al. 2006; Morris et al. 2004).

4.1.1.4 Toxicity

Relatively little is known about the toxic effects of brominated flame retardants in wildlife and man. However, PDBEs are of great concern because of their structural resemblance to polychlorinated biphenyls (PCBs) (Darnerud 2003), well known environmental toxicants. Furthermore, *in vitro* studies using cell culture and *in vivo* studies using laboratory animals show that PBDEs, HBCD and TBBPA all induce toxic effects.

For example, PentaBDE, DecaBDE and HBCD had a negative impact on the thyroid hormone system *in vivo* and TBBPA had negative impacts *in vitro* (Darnerud 2003; Legler and Brouwer 2003). Effects on the thyroid hormone system raise the potential for impacts on growth and development. Legler and Brouwer (2003) noted that the ability of brominated flame retardants to disrupt the thyroid hormone system is evidence that they are potential endocrine (hormone) disruptors and studies also indicate that they may interfere with the oestrogen hormone system. A study on wild grey seal pups (*Halichoerus grypus*) reported that levels of PBDEs in their blubber were statistically linked to levels of thyroid hormones in their blood.

The study suggested that although these results were not sufficient to be sure that PBDEs caused changes in thyroid hormone levels, they were in accordance with the hypothesis that PBDEs are endocrine disruptors (Hall et al. 2003). Research on a pentaBDE commercial mixture reported that it had a hormone disrupting effect on the male reproductive system of laboratory rats and concluded that these chemicals may pose real concerns for adverse effects on human reproductive function and sexual development (Stoker et al. 2005).

Studies on laboratory animals demonstrated that some PBDEs cause toxicity to the nervous system and the immune system (Darnerud 2003; de Wit 2002). Adverse effects of PBDEs on the developing nervous system in mice have been reported (Viberg et al. 2006). In other studies, PentaBDE, DecaBDE and HBCD (de Wit 2002) and TBBPA (Tada et al. 2007) caused changes to the liver.

4.1.1.5   Levels of Brominated Flame Retardants in Marine Wildlife

Brominated flame retardants have been shown to be globally ubiquitous pollutants and contaminate marine wildlife all over the world including in coastal areas, the deep oceans and the Arctic (e.g. de Wit 2002; Law et al. 2003, 2006a; de Wit et al. 2006). The majority of environmental studies have focused on the PBDEs, particularly PentaBDE, and less research has been conducted on HBCD and TBBPA.

PBDEs

Studies on PBDEs in wildlife generally monitor the levels of certain individual PBDE chemicals in tissue. The chemicals that are most commonly studied are those in the PentaBDE formulation, which are mainly BDE-47 and BDE-99 and, to a lesser extent, BDE-17, BDE-28, BDE-153 and BDE-154 (Alaee et al. 2003). Some studies also monitor BDE-183, which is the main constituent of OctaBDE, and BDE-209 which is the primary constituent of DecaBDE. It is not always possible to make direct comparisons of the total levels of PBDEs between different studies because the exact individual PBDE chemicals monitored may differ between studies. Generally, BDE 47 is the most common PBDE found in living organisms (de Wit et al. 2006; Birnbaum and Staskal 2004).

Levels in Fish

One piece of research investigated levels of PBDEs in the muscle tissue of skipjack tuna (*Katsuwonus pelamis*) collected from offshore waters of several countries in 1996–2001 including Japan, Taiwan, Philippines, Indonesia, Seychelles and Brazil, and the Sea of Japan, East China Sea, South China Sea, Indian Ocean and North Pacific Ocean (Ueno et al. 2004). PBDEs were found in almost all samples at levels of <0.1 up to 53 ng/g lipid and indicated the very widespread contamination of the marine environment with PBDEs. Higher levels were apparent in the northern hemisphere compared to the southern hemisphere, which may reflect greater usage of PBDEs in the northern hemisphere. The study noted concern regarding certain Asian countries receiving large amounts of waste electrical equipment and suggested that some developing countries around the East China Sea are potentially 'hotspots' for releasing PBDEs into the marine environment.

Other studies on marine fish include research from 2004 on four species from the North Sea and Northeast Atlantic (cited by Law et al. 2006a). Total PBDEs in fish muscle were 14 ng/g lipid for herring (*Clupea harengus*), 6.7 ng/g lipid for plaice (*Pleuronectes platessa*), 9.7 ng/g lipid for trout (*Salmo trutta*) and 0.4 ng/g lipid for halibut (*Hippoglossus hippoglossus*). A study on fish collected from coastal Florida reported mean concentrations of PBDEs in fish muscle ranging from 8.0 ng/g lipid weight in silver perch (*Bairdiella chrysoura*) to 87.5 ng/g lipid weight in hardhead catfish (*Arius felis*) (Johnson-Restrepo et al. 2005). In three shark species, mean concentrations ranged from 37.8 ng/g lipid in spiny dogfish (*Squalus acanthias*) to 1,630 ng/g lipid in bull sharks (*Carcharhinus leucas*). In sharks, BDE-209 accounted for 58% of the PBDEs detected and its presence suggested exposure to DecaBDE.

In more northerly latitudes, for example southwest Greenland, total PBDEs recorded for Atlantic cod (*Gadhus morhua*) muscle were 480 ng/g lipid. For two sites off Norway, samples of Atlantic cod liver had PBDE levels of 11.7–16.3 ng/g lipid and 15–25 ng/g lipid respectively. BDE-209 was detectable in these samples (studies cited by de Wit et al. 2006).

Levels in Seabirds

Few studies are available on levels of PBDEs in seabirds. Braune and Simon (2004) reported that total PBDEs were detectable in liver samples and eggs of northern fulmars (*Fulmaris glacialis*), black-legged kittiwakes (*Rissa tridactyla*), and thick-billed murres (*Uria lomvia*) collected from Arctic Canada in 1993. Vorkamp et al. (2004b) found that PBDEs were present in black guillemot (*Cepphus grylle*) livers from Greenland. High levels of PBDEs were reported in muscle samples from glaucous gulls (*Larus hyperboreus*) (1,400 ng/g lipid) collected on Bjørnøya (Bear Island, 500 km north of the Norwegian mainland) between 1998 and 2003. BDE-209 was detectable in 30% of blood samples collected from glaucous gulls in 2002 and 2004 (with a mean concentration 410 ng/g lipid) (studies cited by de Wit et al. 2006).

Levels in Marine Mammals

Levels of PBDEs in a number of marine mammals from various locations are presented in Table 4.1. Levels of PBDEs in seals in Arctic regions were an order of magnitude lower than levels found in more southerly regions. For harbour porpoises (*Phocoena phocoena*), two studies showed that concentrations of PBDEs ranged up to particularly high levels in these animals (highest levels 5,800 and

Table 4.1 Levels of PBDEs in marine mammals from different regions of the world

Species	Location	Date of sample collection	Total PBDEs ng/g lipid	Reference
Grey seal (*Halichoerus grypus*) pups	North Sea	1998–2000	45–1,500, Mean 290 (blubber)	Kalantzi et al. (2005)
Harbour seal (*Phoca vitulina*)	San Francisco Bay	1989–1998	Mean 1,730 (blubber)	She et al. (2002)
Ringed seal (*Phoca hispida*)	Northeastern Greenland	Study published 2004	Mean 58 (blubber)	Cited by de Wit et al. (2006)
Ringed seal	Svalbard	Study published 2004	Mean 18	Cited by de Wit et al. (2006)
Ringed seal	Arctic Canada	Study published 2004	4.2–30	Cited by de Wit et al. (2006)
Harbour porpoise (*Phocoena phocoena*)	Iceland, Norway, Baltic Sea and North Sea	1997–2001	18–5,800 (blubber)	Cited by Law et al. (2006a)
Harbour porpoise	England and Wales	1996–2000	Not detectable, up to 7,670 (blubber)	Cited by Law et al. (2003)
Sperm whales (*Physeter macrocephalus*)	Stranded on Netherlands coast	Study published 1998	78–136 ng/g wet weight (blubber)	Cited by Law et al. (2003)
Beluga (*Delphinapterus leucas*)	Svalbard	1998	29–161 (blubber)	Cited by de Wit et al. (2006)
Beluga	Western Canada, Arctic	2001	17 (blubber)	Cited by de Wit et al. (2006)
Long-finned pilot whales (*Globicephala melas*), female	Faroe Islands	Study published 2004	372	Vorkamp et al. (2004a)
Long-finned pilot whales, juvenile	Faroe Islands	Study published 2004	1,018	Vorkamp et al. (2004a)
Polar Bear (*Ursus maritimus*)	Arctic Canada and Alaska	1994–2002	Means 7.6 to 22 (fat)	Muir et al. (2006)
Polar bear	East Greenland	1999–2002	Mean 70 (fat)	Muir et al. (2006)
Polar bear	Svalbard	1999–2002	Mean 50 (fat)	Muir et al. (2006)

7,670 ng/g lipid). PBDEs were present in three sperm whales (*Physeter macrocephalus*) that were found stranded on the coast of the Netherlands. These animals feed in deep offshore waters, implying that PBDEs even contaminate deep-water oceanic food webs.

In Arctic marine mammals, the highest concentrations of PBDEs (372–1,018 ng/g lipid) were recorded in long-finned pilot whales (*Globicephala melas*) lipid, by Vorkamp et al. (2004a). A previous study on pilot whales in the Faroes cited by de Wit et al. (2006) reported even higher levels (61–3,200 ng/g lipid). PBDEs were detectable in polar bears (*Ursus maritimus*) from different regions of the Arctic (Muir et al. 2006). It was also noted that a possible breakdown product (metabolite) of PBDEs was present in polar bears. This indicated that the bears may metabolise PBDEs and therefore the levels of the parent compound PBDEs measured in polar bears may underestimate their total exposure.

Breakdown of DecaBDE in the Environment to Other PBDEs

There is some evidence that once released in the environment, BDE-209, the main constituent of DecaBDE, may degrade to chemicals with a lower degree of bromination that are more toxic and bioaccumulative than BDE-209. This process of degradation is known as debromination and it has been reported to occur in the presence of UV light, by microbial metabolism in sewage sludge and possibly by metabolism in fish and humans (Stapleton 2006).

HBCD

Studies in Europe, Scandinavia and the Arctic have detected HBCD in the tissues of marine invertebrates and fish, seabirds and marine mammals in recent years (see Table 4.2). HBCD was present in muscle and liver samples of several fish species taken in these regions. It was present in guillemot eggs from the Baltic and glaucous gulls from Svalbard, in the Norwegian Arctic.

For marine mammals, comparatively high concentrations of HBCD were present in the blubber of harbour porpoises from Western European seas (up to a median of 5,100 ng/g lipid). In the Arctic, ringed seals from Svalbard (north of mainland Norway) contained higher levels of HBCD than those from Arctic Canada. HBCD was detectable in 100% of polar bear tissues collected from Eastern Greenland and Svalbard and in 13% from Alaska (Muir et al. 2006).

TBBPA

Studies on the levels of TBBPA in the environment are very limited at present. One study detected TBBPA in invertebrates and fish from the North Sea (Morris et al. 2004). For example, levels ranged from 5–96 ng/g lipid in the common whelk (*Buccinium undatum*), <1–35 ng/g lipid in the hermit crab (*Pagurus bernhardus*) and <97–245 ng/g lipid in whiting (*Merlangius merlangus*) muscle. A 2004 publication cited by de Wit et al. (2006) gave TBBPA levels in Atlantic cod livers (*Gadus morhua*) from two sites in Norway of 0.5 and 2.5 ng/g lipid respectively. A study on harbour porpoises (*Phocoena phocoena*) stranded or unintentionally caught in the UK tested 68 samples for TBBPA (Law et al. 2006b). TBBPA was detected in

Table 4.2 Levels of HBCD in marine animals from Europe, Scandinavia and the Arctic

Species	Location	Date of sample collection	Total HBCD ng/g lipid	Reference
Marine invertebrates and fish				
Common whelk (*Buccinium undatum*)	North Sea	1999	29–47 (whole)	Morris et al. (2004)
Cod (*Gadus morhua*)	North Sea	1999	<0.7–50 (liver)	Morris et al. (2004)
Herring (*Clupea harengus*)	Swedish coast	1999–2002	1.5–31 (muscle)	Asplund et al. (2004)
Several fish species	Western Scheldt Estuary, Netherlands	2001	9–1,110 (muscle and liver)	Janák et al. (2005)
Atlantic cod (*Gadus morhua*)	Norwegian Sea	2003	Not detectable, to 51.2 (liver)	Bytingsvik et al. (2004)
Polar cod (*Boreogadus saida*)	Bear Island, Svalbard	2003	7.67–23.4 (liver)	Bytingsvik et al. (2004)
Seabirds				
Glaucous gulls (*Larus hyperboreus*)	Bear Island, Svalbard	2002–2004	6.1–120 (plasma)	Cited by de Wit et al. (2006)
Guillemot (*Uria aalge*)	Baltic Sea	2001	64–220 (eggs)	Sellström et al. (2003)
Marine mammals				
Ringed seal (*Phoca hispida*)	Svalbard	Study published 2004	50–100	Cited by de Wit et al. (2006)
Ringed seal	Canadian Arctic	Study published 2004	1.3–4.7	Cited by de Wit et al. (2006)
Harbour porpoise (*Phocoena phocoena*)	Western European seas	Not given	Median (blubber) Irish Sea 2,900 SW Scotland 5,100 S. Ireland 1,200 S. North Sea 1,100 NW Spain 100	Zegers et al. (2005)
Common dolphin (*Delphinus delphis*)	Western European seas	Not given	Median W. Ireland 900 English Channel 400 NW Spain 200	Zegers et al. (2005)
Beluga (*Delphinapterus leucas*)	Eastern Canadian Arctic	Study published 2004	Means in 2 locations 9.8 and 18	Cited by de Wit et al. (2006)
Polar bear (*Ursus maritimus*)	Eastern Greenland	1999–2002	32.4–58.6	Muir et al. (2006)
Polar bear	Svalbard	1999–2002	18.2–109	Muir et al. (2006)
Polar bear	Bering-Chuckchi Sea (Alaska)	1994–2002	<0.01–35.1	Muir et al. (2006)

18 of the blubber samples at concentrations from 6 to 35 ng/g wet weight. Another study on five harbour porpoises from the UK documented levels of 0.1–418 ng/g lipid in blubber (Morris et al. 2004).

4.1.1.6 Temporal Trends of Brominated Flame Retardants in Marine Wildlife

Trends in levels of PBDEs in marine wildlife over time have been reported in a number of studies. Fewer studies have investigated levels of HBCD in marine wildlife over time, and there appeared to be no research on trends of TBBPA.

PBDEs

Studies have generally shown an increasing trend of PBDE levels in marine wildlife over time, through the 1980s. In the 1990s, studies in some species also reveal an increasing trend while others show a levelling off and sometimes a decreasing trend.

A study on mussels (*Mytilus edulis* and *Mytilus galloprovincialis*) that were collected and freeze-dried from the French coast (English Channel) between 1981 and 2003 monitored levels of PBDEs (Johansson et al. 2006). Results showed that levels increased markedly from 1981 to 1993 doubling about every 5–6 years. Subsequently there was a slow decline up to 2003, except between 1999 and 2001 when levels were higher due to extra inputs of suspended particle matter from intensive flooding and dredging activities.

A study on levels of PBDEs in guillemot (*Uria aalge*) eggs was carried out on eggs that had been collected and stored annually from the Baltic from 1969 to 2001 (Sellström et al. 2003). Levels of PBDEs increased from the 1970s to the 1980s, peaking around the mid- to the late-1980s. Thereafter levels decreased significantly. The decrease in levels may have been due to a voluntary reduction in use of PentaBDE and OctaBDEs by some industries in Europe. A study in Japan on northern fur seals (*Callorhinus ursinus*) also showed a decreasing trend in levels of PBDEs in recent years (Kajiwara et al. 2004). Fat tissues of female northern fur seals were collected at ten time periods between 1972 and 1998 from the Pacific coast of northern Japan. Levels increased from 1972 and peaked around 1991 to 1994. Subsequently, levels decreased to about 50% of the former levels in 1997–1998. It was suggested that the decreasing trend probably resulted from a voluntary phasing out of PentaBDE by Japanese industries in 1990.

Contrary to other European studies on mussels in France and guillemot eggs in the Baltic described above, a study on Atlantic cod (*Gadus morhua*) sampled in Norway showed an increasing trend of PBDE levels in recent years (Bytingsvik et al. 2004). The study sampled cod from an estuary in 1998 and again in 2003. Levels of all BDEs that were monitored (BDE-28, BDE-47, BDE-100, BDE-154) increased by three to four times during the time period.

Studies on marine mammals from USA and Canada have shown no indication of a decrease of PBDEs in recent years. For example, a study on harbour seals (*Phoca vitulina*) from the San Francisco Bay area reported that levels of PBDEs found in stranded seals between 1989 and 1998 increased dramatically (She et al. 2002). A study on ringed seals (*Phoca hispida*) from the Northwest Territories, Canadian Arctic, in 1981, 1991, 1996 and 2000, showed that levels increased exponentially between 1981 and 2000 (cited by Law et al. 2003). Similarly, a study on beluga whales (*Delphinapterus leucas*) from southeast Baffin, Canadian Arctic, in 1982, 1986, 1992 and 1997 showed that PBDE levels increased significantly between 1982 and 1997 (cited by de Wit et al. 2006).

HBCD

Few studies have investigated trends in levels of HBCD in marine wildlife. A study on guillemot (*Uria aalge*) eggs from the Baltic Sea reported an increasing trend in HBCD levels from the mid-1980s to the mid-1990s followed by a levelling out of concentrations up to 2001 (Sellström et al. 2003). For marine mammals, a study on 85 harbour porpoises (*Phocoena phocoena*) stranded or caught unintentionally in the UK between 1994 and 2003 indicated that there was a sharp increase in levels beginning in about 2001 (Law et al. 2006b). It was suggested that the increasing trend in HBCD levels in animals from different areas may reflect a general increase in use and discharge of HBCD due to its use as a replacement for PentaBDE and OctaBDE for some applications within Europe.

4.1.1.7 Policy

From the above review it is clear that PBDEs and HBCD are widespread pollutants in the marine environment that are of concern with respect to their persistence, bioaccumulation potential and toxicity. In a recent review of levels and trends of brominated flame retardants in the Arctic environment, de Wit et al. (2006) wrote:

> We conclude that Penta-, Octa- and DecaBDE, PBBs and HBCD have characteristics that
> qualify them as POPs according to the Stockholm Convention.

International Policy

The Stockholm Convention is a global treaty which necessitates action by all governments to eliminate or reduce POPs listed by the Convention. While the list does not currently include any brominated flame retardants, OctaBDE has been proposed for the list and PentaBDE is under review for inclusion (Earth Negotiations Bulletin 2006). This is encouraging given the global contamination and nature of these chemicals. However, no action has been taken on DecaBDE, HBCD or TBBPA under this Treaty. Given their occurrence in living organisms and their persistence and potential toxicity, it is of great concern that these chemicals have not yet been proposed for international action under the Stockholm Convention.

Fortunately, some regional and national policies have been adopted which aim to eliminate or reduce the use of brominated flame retardants.

Regional Policy

Brominated flame retardants have been considered by The Convention for the Protection of the Marine Environment of the North-East Atlantic (known as the 'OSPAR Convention'). In 1998, the Ministerial Meeting of OSPAR agreed on the target of cessation of discharges, emissions and losses of all hazardous substances to the marine environment by 2020 (the 'one generation' cessation target, OSPAR 1998a, b). PBDEs, HBCD and TBBPA are included on the OSPAR list of chemicals for priority action towards this target (OSPAR 2006).

PentaBDE and OctaBDE technical products were banned in the European Union after August 2004, following the outcome of risk assessments. For DecaBDE and HBCD, EU risk assessments are still ongoing but as yet they have not recommended the reduction of these chemicals (ENDS 2007). For TBBPA, a risk assessment report in 2006 (EU 2006) concluded that "no health effects of concern have been identified for TBBPA", though possible risks have since been identified for the aquatic environment.

Also within Europe, the restriction of the use of certain hazardous substances in electrical and electronic equipment, the RoHS Directive (EC 2003c), prohibits the use of PBDEs in greater than regulated amounts.

National Policy

Bans for PentaBDE and OctaBDE products are set to become effective in several states of the USA (de Wit et al. 2006). In Japan, PentaBDE was voluntarily withdrawn from the market (Vorkamp et al. 2004a). Sweden has passed legislation that restricts the use of DecaBDE in new products in sectors such as textiles, upholstery and electrical wiring as of January 2007 (ENDS Europe Daily 2006). The government of Norway has banned Deca-BDE in electronic and electrical products, and a ban on Deca-BDE in textiles, furniture filling and cables will take effect during 2008 (ENDS Europe Daily 2008).

4.1.1.8 Conclusion

PentaBDE, OctaBDE, DecaBDE and HBCD are all persistent global pollutants. While regional bans on PentaBDE and OctaBDE are a step in the right direction, ultimately action at a global level, supported by their inclusion under the Stockholm Convention, will be needed to prevent their continued use and release into the environment. Deca-BDE and HBCD are still in widespread use but it is known that they also have the characteristics of POPs. It is therefore essential that they are also included by the Stockholm Convention and are phased out of use. There has been limited research on TBBPA but it has nevertheless been reported to contaminate some species of marine organisms, and has toxic properties. It would therefore be prudent to take a precautionary approach with respect to this chemical, and to reduce any further potential toxic impacts on living organisms by phasing out its use on a global scale.

Unfortunately, even if the brominated flame retardants are eventually phased out globally they will leave a legacy for years to come as they continue to leach out from materials in which they have been incorporated, and because of their longevity in the environment.

4.1.2 Perfluorinated Compounds (PFCs)

Perfluorinated compounds (PFCs) are man-made chemicals that never occur in nature. They have been produced industrially for over 50 years. PFCs have properties of repelling both water and oil. Consequently, they have been widely used as protective coatings for a variety of products and they are also used as surfactants (to reduce surface tension). However, PFCs are persistent in the environment and some are known to bioaccumulate, building up in the blood and liver of living organisms. Testing of some PFCs also shows that these chemicals have toxic effects in animals. Unfortunately, PFCs have become widespread environmental pollutants and even contaminate the tissues of living organisms in remote regions such as the Arctic. Their properties of persistence, bioaccumulation, toxicity and in some cases long range transport on air currents, indicate that they are problematic and of very high concern (Allsopp et al. 2005; Giesy and Kannan 2002).

PFCs are composed of a carbon-fluorine chain and generally have side groups attached such as sulphonic acids or carboxylic acids. These compounds are respectively called perfluorinated sulphonates, or perfluoroalkyl sulphonates and perfluorinated carboxylates, or perfluoroalkyl carboxylates (PFCA). Together they represent two major classes of PFCs (Giesy and Kannan 2002).

There are two industrial processes used to make PFCs. One was used by the 3 M company until 2003 to produce perfluoroalkyl sulphonates and is known as electrochemical fluorination (Hekster et al. 2003). One chemical which was generated by this process is perfluorooctane sulphonate (PFOS), now a ubiquitous environmental contaminant. The other process used to produce PFCs is the telomerization process. Perfluorooctanoic acid (PFOA) and fluorotelomer alcohols are some of the chemicals produced (Dinglasan et al. 2004; Hekster et al. 2003; Renner 2001). Fluorotelomer alcohols have been shown to break down in the environment to form perfluorinated carboxylates, chemicals that contaminate body tissues of some species (see below).

The carbon-fluorine bond in PFCs is very strong and gives thermal and chemical stability to many PFCs (So et al. 2004). The stability that makes fluorinated compounds desirable for commercial use also makes them potentially significant environmental contaminants due to their resistance to natural breakdown processes; that is, their persistence (Key et al. 1997).

4.1.2.1   Uses of PFCs

PFCs are manufactured because of their specific physical and chemical properties such as chemical and thermal inertness and special surface-active properties (Hekster et al. 2003). They repel both water and oil and act as surfactants, that is, they reduce surface tension, and do so better than other surfactants (Renner 2001). These properties have led to the use of PFCs in a wide variety of applications. For instance, their unique properties of repelling both water and oil have led to the use of perfluroalkylated substances as coatings for carpet protection, textile protection, leather protection, and paper and board protection. They are also used in firefighting foams and as polymerization aids. In addition they are used as speciality surfactants, for example, in cosmetics, electronics, etching, medical use and plastics (Hekster et al. 2003; So et al. 2004).

A specific use of the chemical PFOA is in the manufacture of Teflon, a nonstick coating used for saucepans (ENDS 2003). PFOA is used also used to make Goretex (Renner 2003). Fluorotelomer alcohols are used as precursor molecules for the production of fluorinated polymers, which, in turn, have similar uses to PFOS-based compounds such as in paper and carpet treatments. Fluorotelomer alcohols are also used in the manufacture of paints, adhesives, waxes, polishes, metals and electronics (Dinglasan et al. 2004).

4.1.2.2   Sources of Environmental Contamination

The mechanisms and pathways leading to the presence of perfluorinated compounds in wildlife and humans are not well characterised, but it is likely there are multiple sources (Kannan et al. 2002a). The mechanisms by which volatile PFCs enter the environment may include release during manufacturing and application processes (Stock et al. 2004). For example, it has been demonstrated that the effluent from a fluorochemical manufacturing facility can be a source of perfluoroalkylated substances in the environment (Hekster et al. 2003). Emissions to the environment may also occur by leaching from consumer products and from waste products in landfills, due to biological or non-biological degradation processes (Stock et al. 2004). Firefighting foams can result in direct releases of PFCs to the environment

(Moody and Field 1999; Schultz et al. 2004). For example, at a fire in an oil depot in Buncefield, UK, in December 2005, vast quantities of PFOS-foam were used to contain the fire, which resulted in the local groundwater becoming contaminated with PFOS (McSmith 2006).

PFOS and PFOA are widespread environmental contaminants but, because of their low volatility, these chemicals are unlikely to enter the atmosphere directly and undergo long-range transport on air currents. Instead, it is thought that other more volatile PFCs such as perfluoroalkyl sulphonamides, are transported for long distances on air currents and are subsequently degraded in the environment or metabolised by living organisms to form PFOS (Shoeib et al. 2004; So et al. 2004; Stock et al. 2004). Research shows that PFCs produced by electrochemical fluorination can be broken down by microorganisms to PFOS and PFOA (Hekster et al. 2003).

Some perfluoroalkyl carboxylates have been found to contaminate the tissues of many living organisms and yet they have not been widely used in consumer or industrial materials. Their widespread occurrence in the environment therefore suggests that these compounds are also formed as degradation products of precursor chemicals. Plausible precursor chemicals that could break down to yield perfluoroalkyl carboxylates are fluorotelomer alcohols (Ellis et al. 2004). Recent studies have shown that fluorotelomer alcohols can be broken down in the atmosphere to form perfluoroalkyl carboxylates (Ellis et al. 2004) and by living organisms to yield perfluoroalkyl carboxylates (Dinglasan et al. 2004). Studies also show that fluorotelomer alcohols are ubiquitous in the atmosphere of North America (Stock et al. 2004) and that they are capable of long-range transport in the air (Ellis et al. 2003). Ellis et al. (2004) noted that atmospheric degradation of fluorotelomer alcohols is likely to contribute to the widespread dissemination of perfluoroalkyl carboxylates in the environment. Furthermore, the pattern of the different perfluoroalkyl carboxylates that are formed from the breakdown of fluorotelomer alcohols could account for the distinct contamination profile of perfluoroalkyl carboxylates found in Arctic animals. In addition, analysis of the perfluoroalkyl carboxylates present in polar bear liver suggested that the sole input of one perfluoroalkyl carboxylate, namely perfluorononanoic acid (PFNA), was attributed to fluorotelomer alcohols as the source. Thus, long-range transport and degradation of fluorotelomer alcohols could explain the presence of long-chain perfluoroalkyl carboxylates in Arctic animals.

4.1.2.3 Bioaccumulation

Bioaccumulative substances are of great concern because of their potential to attain toxicologically significant tissue and organ residue concentrations in wildlife and humans (Kelly et al. 2004). Unlike many persistent and bioaccumulative environmental pollutants, PFOS and other PFCs do not accumulate in lipids of the body but in the blood and in the liver and gallbladder (Renner 2001; Martin et al. 2003).

In one study on bioaccumulation of PFCs in fish, rainbow trout (*Oncorhynchus mykiss*) were exposed to perfluorinated sulphonates and carboxylates in the surrounding water (Martin et al. 2003). Following exposure, PFC concentrations were found to be greatest in the blood of the fish, followed by the kidney then the liver, then gallbladder, then adipose tissue and finally muscle tissue. Perfluorinated carboxylates with a carbon-fluorine chain of more than six and perfluorinated sulphonates with a chain length of more than four were found to bioaccumulate in the blood and liver of the rainbow trout.

The concentration of some environmental pollutants in animal tissues increases through the food chain, such that the highest levels are found in top predators at the apex of food chains. This process is known as biomagnification. One study conducted on the food web of the bottlenose dolphin (*Tursiops truncatus*) in the USA reported that biomagnification of both PFOS and some long-chain perfluorinated carboxylates occurred through this food web (Houde et al. 2006a).

A study on harbour porpoises (*Phocoena phocoena*) from Northern Europe analysed the concentrations of PFOS in liver tissue of an unintentionally caught mother porpoise and her foetus (Van de Vijver et al. 2004). Levels of PFOS were over twofold greater in the foetus than its mother, which suggested that placental transfer and foetal accumulation of PFOS occurs in porpoises, whales and dolphins (Houde et al. 2006b).

4.1.2.4 Toxicity

PFOS and PFOA have been found to cause a wide range of toxic effects in laboratory rats exposed at doses higher than those currently found in wildlife. For instance, PFOS and PFOA caused toxic effects on the liver, including reduced serum cholesterol levels and induction of enzymes associated with β-oxidation of fatty acids (Berthiaume and Wallace 2002). PFOA caused adverse effects on the immune system (Yang et al. 2001, 2002) and acted as a tumour promoter (Adinehzadeh et al. 1999). PFOS caused adverse effects on development, including the death of rat pups (Lau et al. 2003, 2004). At levels lower than those present in wildlife, PFOS was found to act as an endocrine disruptor in rats, affecting the oestrous cycle (Austin et al. 2003).

Some PFCs have been found to inhibit the communication system between cells in both mammalian cell lines *in vitro* and in rats. Disruption of this process results in abnormal cell growth and function and, as such, the PFCs may pose a risk to the health of mammalian systems (Hu et al. 2002).

Southern sea otters (*Enhydra lutris nereis*) from the Californian coast are reported to be affected by diseases and high mortality rates. The causes of this are likely to be multifactorial and possibly include habitat destruction, pollutants, climate change, and over-harvesting of marine resources. One study investigated whether levels of certain PFCs in the sea otters were linked to disease in the animals (Kannan et al. 2006). Analysis of results showed that concentrations of both PFOS and PFOA were significantly higher in sea otters with infectious disease. It is not known from the results whether the elevated levels of PFOS and PFOA were a cause of increased disease but the fact that there was an association between higher levels and disease is a cause for concern and suggested the need for further studies.

4.1.2.5   Levels of PFCs in Marine Wildlife

Perfluoroalkyl sulphonamides have been detected in the atmosphere (Stock et al. 2004; Shoeib et al. 2004). In the marine environment, PFOS and PFOA were detectable in seawater from coastal areas and open ocean waters of the Pacific and Atlantic Oceans (Yamashita et al. 2005; Taniyasu et al. 2004). PFOA and PFOS were even detectable in trace quantities in samples collected from the open ocean at depths of greater than 1,000 m (Taniyasu et al. 2004). Yamashita et al. (2005) reported that levels of PFOS and PFOA in coastal regions of Asia were greater than levels in offshore waters in the Pacific and Atlantic, suggesting sources associated with urban and industrial areas.

With regard to marine wildlife, research has shown that some PFCs contaminate the body tissues of many species including fish, birds and marine mammals.

Marine Fish

A study of 20 species of fish from different coastal regions of Japan found PFOS concentrations ranged from 1 to 489 ng/ml in blood and 3 to 7,900 ng/g wet weight in liver (Taniyasu et al. 2003). A study of fish from the Italian coast of the Mediterranean analysed bluefin tuna (*Thunnus thynnus*), and swordfish (*Xiphias gladius*) samples and found PFOS concentrations in blood ranged from 4 to 52 ng/ml and in liver ranged from <1 to 87 ng/g wet weight (Kannan et al. 2002a). Perfluorooctane sulphonamide (FOSA) was also detectable in the blood of the fish. A study of PFOS in oysters (*Crassostrea virginica*) from the Gulf of Mexico reported concentrations ranging from <42 to 1,225 ng/g dry weight (Kannan et al. 2002b).

Seabirds

A study of common cormorants (*Phalacrocorax carbo*) from the coast of Sardinia, Italy, recorded PFOS in liver tissue at concentrations ranging from 32 to 150 ng/g wet weight (Kannan et al. 2002a). Also detectable in the samples was PFOA, at concentrations of 29–450 ng/g wet weight. Another study reported that levels of PFOS in the livers of common cormorants from Japan were sixfold greater than in cormorants from Sardinia (Kannan et al. 2002c).

A study on the Laysan albatross (*Diomedea immutabilis*) and black-footed albatross (*Diomedea nigripes*) from Midway Atoll, North Pacific Ocean, detected PFOS in blood serum samples at concentrations of 3–34 ng/ml (Kannan et al. 2001a). The authors commented that this result suggested the widespread distribution of sulphonated PFCs in remote marine locations.

Marine Mammals

PFOS has been found to contaminate the liver tissue of many species of marine mammals from a wide range of geographical regions in the Northern Hemisphere, including remote regions such as the Arctic (see Table 4.3).

Table 4.3 Levels in PFOS in the livers of marine mammals

Location	Species	Mean and/or (range) PFOS in liver (ng/g wet weight)	Reference
Northern Europe (Norway, Denmark, Iceland, German Baltic Sea)	Harbour porpoise (*Phocoena phocoena*) (n = 41)	(26–1,149)	Van de Vijver et al. (2004)
North Sea (samples from animals found stranded on Belgian, French and Dutch coasts)	Grey seal (*Halichoerus grypus*) (n = 6)	(11–233)	Van de Vijver et al. (2003)
	Harbour seal (*Phoca vitulina*) (n = 24)	(<10–532)	Van de Vijver et al. (2003)
	Harbour porpoise (*Phocoena phocoena*) (n = 48)	(12–395)	Van de Vijver et al. (2003)
	White beaked dolphin (*Lagenorhynchus albirostris*) (n = 7)	(14–443)	Van de Vijver et al. (2003)
	White-sided dolphin (*Lagenorhynchus acutus*) (n = 2)	(<10–26)	Van de Vijver et al. (2003)
	Sperm whale (*Physeter macrocephalus*) (n = 6)	(19–52)	Van de Vijver et al. (2003)
Baltic Sea	Ringed seal (*Phoca hispida*) (n = 25)	(130–1,100)	Kannan et al. (2002a)
	Grey seal (*Halichoerus grypus*) (n = 27)	(140–360)	Kannan et al. (2002a)
East Coast USA (Florida)	Striped dolphin (*Stenella coeruleoalba*) (n = 2)	212 (36.6–388)	Kannan et al. (2001b)
	Rough-toothed dolphin (*Steno bredanensis*) (n = 2)	54.2 (42.8–65.6)	Kannan et al. (2001b)
	Bottlenose dolphin (*Tursiops truncatus*) (n = 20)	489 (48.2–1,520)	Kannan et al. (2001b)
	Short-snouted spinner dolphin (*Stenella clymene*) (n = 3)	123 (78.7–168)	Kannan et al. (2001b)
	Pygmy sperm whale (*Kogia breviceps*) (n = 2)	14.8 (6.6–23)	Kannan et al. (2001b)
West Coast USA	California sea lion (*Zalophus californianus*) (n = 6)	26.6 (4.6–49.4)	Kannan et al. (2001b)
	Elephant seal (*Mirounga augustirostris*) (n = 5)	9.3 (<5–9.8)	Kannan et al. (2001b)
	Harbour seal (*Phoca vitulina*) (n = 3)	27.1 (10.3–57.1)	Kannan et al. (2001b)
Alaska	Northern fur seal (*Callorhinus ursinus*) (n = 13)	(<10–122)	Kannan et al. (2001b)
	Polar bear (n = 17) (*Ursus maritimus*)	350 (175–678)	Kannan et al. (2001b)
Canadian Arctic	Polar bear (*Ursus maritimus*) (n = 7)	3,100	Martin et al. (2004a)

Table 4.3 shows that, in samples of seal liver from the North Sea and Baltic Sea, the highest levels of PFOS were present in ringed seals (*Phoca hispida*) from the Baltic (130–1,100 ng/g wet weight). In blood samples taken from ringed seals in the Baltic in 1998 (mean concentration 242 ng/ml), levels of PFOS were an order of magnitude greater than blood samples taken from seals at Spitzbergen, in the Norwegian Arctic (mean 10.1 ng/ml) (Kannan et al. 2001b). This result is most likely explained by the fact that the Baltic has industrialised regions which are closer to sources of PFCs than the Arctic.

Levels of PFOS in the livers of harbour porpoises (*Phocoena phocoena*) from Northern European seas ranged up to comparatively high levels (26–1,149 ng/g wet weight) (Van de Vijver et al. 2004). Similarly, in a number of species of dolphins from the North Sea and Florida, levels of PFOS in liver tissue were in the range of tens of nanograms per gram wet weight to over a thousand nanograms per gram wet weight.

A study on marine mammals in the North Sea (Van de Vijver et al. 2003; see Table 4.3) noted that higher levels of PFOS were present in species which fed by the coast (grey seals, harbour seals, harbour porpoise and white beaked dolphin) compared to species that fed further offshore (sperm whale, *Physeter macrocephalus*, and white-sided dolphin, *Lagenorhynchus acutus*). Higher concentrations in coastal species was likely to be due to the closer proximity to PFC sources. Nevertheless, the presence of PFOS in sperm whales suggested that PFCs have even reached deep areas of the ocean.

PFOS was detected in polar bears (*Ursus maritimus*) from the Arctic at comparatively high levels, which most probably reflects their position in the food chain as top predators. Concentrations were about tenfold greater in bears from the Canadian Arctic (1,700–>4,000 ng/g, mean 3,100 ng/g) (Martin et al. 2004a) compared to Alaska (175–678 ng/g, mean 350 ng/g) (Kannan et al. 2001b). This was perhaps because bears from the Canadian Arctic were at a lower latitude than bears from Alaska and were therefore closer to regional sources of PFOS. PFOS was also present in liver samples from polar bears in East Greenland (911–2,140 ng/g) and Svalbard (756–1,290 ng/g) (Smithwick et al. 2005).

In addition to PFOS, a number of perfluorinated carboxylates were also detected in some harbour porpoises from seas of Northern Europe (Van de Vijver et al. 2004) and in polar bears from the Canadian Arctic (Martin et al. 2004a). With regard to PFOA, southern sea otters (*Enhydra lutris nereis*) from the Californian coast were found to have high levels of PFOA in liver tissue (<5–147 ng/g wet weight) (Kannan et al. 2006). The study suggested that the results implied the existence of specific sources of PFOA in coastal California.

4.1.2.6 Temporal Trends in Levels of PFCs in Marine Wildlife

Several studies have investigated trends in levels of PFOS in marine animals over time. One study monitored trends in the levels of PFOS in guillemot (*Uria aalge*) eggs from the Baltic using an archive of frozen eggs (Holmström et al. 2005). PFOS was measured in eggs dating from 1968 to 2003. The results showed that there was an almost 30-fold increase in PFOS concentrations during this time period. For instance, in 1968 the concentration from pooled eggs was 25 ng/g wet weight and by 2000 the concentration was 871 ng/g. Analysis of figures showed that there was a statistically significant increasing trend in levels over time, with a general increase of between 7% and 11% per year. Levels peaked during 1997–2000 and then decreased somewhat in 2001 and 2003. A study on birds from the Arctic, thick-billed murres (*Uria lomvia*), reported an increasing trend in the levels of PFOS between 1987 and 2003 (cited by Houde et al. 2006b).

Levels of PFOS in liver tissue from ringed seals (*Phoca hispida*) were investigated in four different years between 1986 to 2003 in East Greenland, and 1982 and 2003 in West Greenland (Bossi et al. 2005). The study indicated there was a significant increase in levels of PFOS over time in both locations.

A study on levels of PFCs in liver tissue of southern sea otters (*Enhydra lutris nereis*) from the California coast between 1992 and 2002 was conducted. It reported that concentrations of PFOA increased significantly over the time period (Kannan et al. 2006). PFOS concentrations also increased from 1992 to 1998 but then decreased after 2000. It was noted that the decrease in PFOS levels coincided with the phase-out of PFOS-based fluorochemicals by the 3 M company.

A study on temporal trends of PFCs in polar bears (*Ursus maritimus*) from two locations in the North American Arctic was carried out using archived liver tissue samples collected between 1972 and 2002 (Smithwick et al. 2006). The levels of PFOS and certain long-chained perfluorinated carboxylates showed an exponential increase between 1972 and 2002 at both locations.

Figure 4.1 shows a graphic representation of temporal trends of PFOS in several northern species and in humans over the past 3 decades. There are clearly increasing trends for all species.

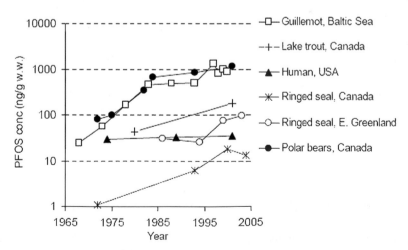

Fig. 4.1 Temporal trend of PFOS (ng/g ww) in guillemot eggs, lake trout homogenates, human serum and livers of polar bears/ringed seals from Arctic Canada and ringed seals from Greenland (Houde et al. 2006a)

4.1.2.7   Current Policy on PFCs

United States

In 2000, the 3 M company made a decision to stop the manufacture of PFOS-related chemicals by 2003 (see Section 4.1.2). It later transpired that the decision to phase out 'perfluorooctanyl chemistry' by 3 M was prompted by pressure from the US Environmental Protection Agency (EPA). Research was accumulating on the presence of PFCs in human blood from around the world and on adverse effects in laboratory animals, and the EPA threatened 3 M with regulatory action if it did not stop the perfluorooctanyl chemistry voluntarily (ENDS 2001, 2004). The US EPA has since imposed a ban on the use of PFOS, with only a few essential uses in the aviation, photographic and microelectronics industries being exempted (ENDS 2004).

The US EPA has also its turned attention to the perfluorinated carboxylate breakdown product and manufacturing aid PFOA (Renner 2004). It produced a draft risk assessment of PFOA in April 2003 (ENDS 2004) in which it identified 'potential human health concerns' (ENDS 2003). More recently, a scientific panel which advises the EPA was given the task of reviewing the EPA's risk assessment of PFOA's health effects. It concluded that the evidence supported classification of PFOA as a 'likely' carcinogen rather than the classification of "suggestive of carcinogenicity" which the EPA risk assessment had proposed (ENDS 2006; US EPA 2006a). Other action on PFOA by the EPA was taken in January 2006, when the EPA launched a global stewardship programme inviting companies to reduce PFOA releases and its presence in products by 95% by no later than 2010. They must also work to eliminate releases by 2015. Fluorinated chemicals that break down to PFOA in the environment were also included together with 'higher homologues', similar fluorinated compounds with longer carbon chains than the eight-carbon PFOA (ENDS 2006; US EPA 2006b).

Europe

In December 2005, the European Commission released a proposal for a directive relating to restrictions on the marketing and use of perfluorooctane sulphonates (Commission of the European Communities 2005). The proposed directive outlines a risk reduction strategy which recommended marketing and use restrictions for PFOS. However, several uses will be permitted to continue indefinitely under the proposed directive. In contrast, national policies for restrictions of PFOS and related substances developed by Sweden (G/TBT Notification 2005) and the UK (DEFRA 2005a) recognise the desirability and feasibility of eventual phase-out of these uses through a series of application-specific deadlines from 2007 to 2010 (with the exception of hydraulic fluids used in aviation). In this regard, Greenpeace has proposed that the European Commission could go much further than the proposed directive by introducing similar phase-out deadlines for currently permitted applications (Greenpeace 2006d). In addition to this, there are environmental and health concerns relating to PFOA and related compounds, and these PFCs are not addressed by the proposed directive. Greenpeace proposes that the European Commission will ultimately need to address a much wider range of substances if it is effectively to protect the environment and peoples of Europe from the threat of PFCs.

In 2005, Sweden called for a global ban on PFOS and made a proposal for PFOS to be considered for inclusion in the Stockholm Convention (Swedish Chemicals Inspectorate KemI 2005). Subsequently, in 2006, a risk profile for PFOS was adopted, a further step towards its inclusion on the Stockholm Convention POPs list (Earth Negotiations Bulletin 2006).

4.1.2.8 Conclusions

PFCs are persistent chemicals, some of which are found in the tissues of marine organisms worldwide. In particular, PFOS, PFOA and certain perfluorinated carboxylates are known to contaminate marine wildlife. In this regard, these chemicals, and PFCs which degrade to these chemicals are of great concern. Greenpeace therefore advocates that all of these PFCs should be nominated for listing by the Stockholm Convention, from where their global phase-out can be prescribed (Greenpeace 2006d). It is important to recognise that the only sustainable strategy to protect the environment and human health from PFCs will be to replace them progressively with less hazardous, preferably non-hazardous, alternatives wherever and whenever they are available (the so-called substitution principle). Of course, in implementing such an approach, it will be essential to ensure that one problematic chemical is not simply replaced with another, with an environmental fate perhaps even less well understood. In this context, it is of concern that the majority of products currently marketed as alternatives to existing PFCs also rely on perfluorinated chemistry (see, for example, Walters and Santillo 2006), a testament to the somewhat unique properties that this basic structure confers. Hence, as part of the substitution process, it will undoubtedly also be vital to consider the fundamental necessity for the diversity of perfluorinated chemicals and materials on which we have increasingly come to rely. In other words, are disposable greaseproof hamburger wrappers, microwavable popcorn bags and stain-resistant carpets truly essential to support our quality of life? Or could we live without them? (Santillo et al. 2006).

4.1.3   The Beginning of the End for Chemical Pollution?

The polybrominated and perfluorinated chemicals discussed here are just two of the diverse array of man-made chemicals that present widespread and long-term threats to marine ecosystems. While the inclusion of key examples of these groups (such as penta- and octa-BDEs and PFOS) under the Stockholm Convention will be welcome and may ultimately lead to world-wide action, the process is slow and addresses only a fraction of the overall problem of marine pollution. On the regional level of the European Union, it has been accepted for some years that standard risk assessment techniques simply cannot be applied to substances in the marine environment which are environmentally persistent or which can build up in body tissues even through low-level exposures, because 'safe' doses of such chemicals cannot be reliably determined. As a result, when the EU Technical Guidance Document for risk assessment was updated in 2003, it included a so-called 'PBT assessment' (for substances which are Persistent, Bioaccumulative and Toxic, of which POPs are a sub-set), placing emphasis on avoiding inputs at source rather than on trying to assess and manage risks in the environment (EC 2003a).

But official recognition of the need for more precautionary approaches to the protection of the marine environment from chemical pollution began well before this, exemplified by the ground-breaking target of cessation of releases of hazardous substances to the North Sea within one generation, agreed by Ministers of many northern European states at the 4th North Sea Conference held in Esbjerg (Denmark) in 1995 (4th NSC 1995). This cessation target was later formalised under the OSPAR Convention (for the Protection of the Marine Environment of the North-East Atlantic). At its Ministerial Meeting in 1998, OSPAR adopted a strategy to achieve the cessation of discharges, emissions and losses of hazardous substances which reach or could reach the sea, by 2020 (OSPAR 1998a, b). Since then, the list of hazardous substances or groups of substances identified by OSPAR for priority action has grown from 12 to more than 40 (OSPAR 2006), including a number of chemicals which remain in major commercial use.

Despite the aspirations, however, progress towards the cessation target has so far been slow and difficult to measure. This has resulted, in part, from a redirection of national efforts to the development of a whole new system of regulations governing the manufacture, marketing and use of chemicals right across Europe, a system finally agreed in December 2006 (EC 2006). Under the new regulations, termed REACH (Registration, Evaluation and Authorisation or CHemicals), the burden of proof of chemical safety has been shifted significantly from governments to industry, with companies needing to seek specific authorisation for the continued use of some of the most hazardous chemicals, including PBT substances. In particular, REACH introduces for the first time a requirement for the substitution of many of these most hazardous chemicals, when safer alternatives are available.

Although it remains uncertain how effective REACH will be in practice, it is without doubt a significant step forward. Whether it provides sufficient tools to meet OSPAR's cessation target and thereby protect the marine environment from chemical pollution, and whether it can provide a model for more widespread controls on the use of hazardous chemicals, remain to be seen.

4.2   Radioactive Pollution

Artificial radionuclides, produced as a result of nuclear weapons testing and waste from nuclear power stations, now pollute the oceans worldwide. Many artificial radionuclides have no natural counterparts and have extremely long lifetimes

(Johnston et al. 2000). Pollution of the environment by artificial radionuclides is of concern because they can act as potent carcinogens and mutagens (Cockerham and Cockerham 1994).

Nuclear weapons testing, predominantly between 1954 and 1962, has been the largest single source of artificial radionuclides to the oceans, due to fallout. Other sources arise from the dumping of radioactive waste at sea and operational discharges from nuclear power facilities and nuclear reprocessing plants (Johnston et al. 1998). Presently, the most prominent point sources of radioactive pollution to the oceans are from the Sellafield (UK) and La Hague (France) nuclear reprocessing plants (Johnston et al. 2000). This section briefly discusses contamination of the marine environment resulting from the Sellafield plant. Radioactive pollution resulting from the use of nuclear power is an ongoing problem and, as yet, there are no plans for a global phase-out of its use.

Spent uranium fuel from nuclear reactors contains plutonium, uranium and other fission products (high level nuclear waste). Reprocessing of spent fuel is a technique that chemically separates plutonium and uranium from high level nuclear waste. The process was originally developed for extracting plutonium for use in nuclear weapons. It was later intended that the plutonium could be used as fuel in breeder reactors but this technology did not take off. Consequently, most reprocessed fuel remains in storage. It is the opinion of Greenpeace that reprocessing of spent nuclear fuel is therefore illogical because it produces useless materials and increases nuclear proliferation risks. In addition, reprocessing results in the discharge of significant amounts of radioactively contaminated liquid waste to the oceans (Greenpeace 2006b). In 1998, Greenpeace found that sediment from the seabed near Sellafield was so contaminated it should be classified as nuclear waste (Greenpeace 1999). Furthermore, the polluting footprint of Sellafield is not only localised but is widespread, and has led to radioactive contamination from the Irish Sea to waters of the Arctic. This is due to the long-distance transport of radionuclides on ocean currents.

In the 1970s and 1980s, discharges of the radionuclide caesium-137 (^{137}Cs) contributed to the most obvious radioactive pollution from Sellafield to the North Atlantic and the Arctic. In addition, significant quantities of ^{134}Cs and strontium-90 (^{90}Sr) have been transported to the Arctic together with measurable quantities of plutonium and americium. The introduction of improved technologies has led to a reduction in the release of these radionuclides discharged from Sellafield. However, the introduction of newer operations has also led to contamination from other radionuclides. From 1995 onwards, the operation of the Thermal Oxide Reprocessing Plant (THORP) has resulted in increased amounts of iodine-129 (^{129}I) discharges. From 1994 onwards, the Enhanced Actinide Removal Plant (EARP) has resulted in discharges of Technetium-99 (^{99}Tc) (Kershaw et al. 1999).

With regard to ^{137}Cs, reprocessing at Sellafield between 1952 and 1998 led to discharges to the Irish Sea with discharges peaking in 1974–1978 (Osvath et al. 2001). A survey of sediments in the Irish Sea by Osvarth et al. (2001) showed contamination by ^{137}Cs in a strip extending northwards along the coast from the discharge point. Research has shown that Sellafield has been an important source of ^{137}Cs in the Arctic Ocean due to discharges being transported with the Norwegian Coastal Current from the Northeast Atlantic to the Arctic Ocean (Aarkrog et al. 2000). It is estimated that greater than 90% of the total ^{137}Cs activity discharged from Sellafield has been removed from the Irish Sea by the prevailing water circulation. However, despite the fact that ^{137}Cs is highly soluble in ocean waters, there has also been significant incorporation of this radionuclide into the seabed of the Irish Sea. Consequently, the remobilisation of contaminated sediments in this area presently acts as a continued source of ^{137}Cs to waters of the Irish Sea (McCubbin et al. 2006). A study of dead stranded grey seals (*Halichoerus grypus*), harbour seals (*Phoca vitulina*) and harbour porpoises (*Phocoena phocoena*) found around the UK coast detected ^{137}Cs in tissue samples from the animals (Watson et al. 1999). The levels of contamination in the seals and porpoises decreased with increasing distance from Sellafield, which indicated that Sellafield was the major source of their tissue contamination. These marine mammals concentrated radiocaesium from their environment at levels of 300 times greater than the level in seawater.

Plutonium discharges from Sellafield peaked in the early 1980s. Most of the plutonium and americium discharged from Sellafield has been incorporated in the subtidal sediments of the Irish Sea. However, plutonium is also known to have been transported over 2,500 km to the Arctic. Plutonium, which resides in sediment, presently acts as a source to overlying waters from where it continues to be transported to the Arctic (Kershaw et al. 1999). Plutonium originating from Sellafield was found to contaminate seaweed (*Fucus vesiculosus*) collected from the western Irish coastline between 1986 and 1996 (Ryan et al. 1999). The same study also showed that plutonium contaminated mussels (*Mytilus edulis*) and oysters (*Crassostrea gigas*) on the northeast coast of Ireland and there was no decline in radioactivity levels between the early 1990s and 1997s.

Technetium-99 (^{99}Tc) has a half-life of 213,000 years (Lindahl et al. 2003). Enhanced releases of ^{99}Tc from Sellafield occurred between 1994 and 2002 and since then discharge has been reduced (McCubbin et al. 2006). [99]Tc has been found to contaminate sediments of the Irish Sea, with activity decreasing with increasing distance from Sellafield. It also contaminates the Norway lobster (*Nephrops norvegicus*) that is fished in the Irish Sea (McCubbin et al. 2006). Research has shown that ^{99}Tc from Sellafield contaminates waters north of the Irish Sea. For example, studies of seaweed (*Fucus vesiculosus* and *Fucus serratus*) on the Swedish west coast showed that pulses of ^{99}Tc discharged from Sellafield in the mid-1990s could be detected in seaweeds 4–5 years later (Lindahl et al. 2003). Similarly, pulses from 1995 were detectable in seawater by 1997 in northern Norway (Kershaw et al. 1999).

It is clear from the above discussion that discharges of radioactive waste from Sellafield have led to contamination of the Irish Sea and oceanic waters stretching as far north as the Arctic. Continuing discharges of liquid radioactive waste to the oceans from Sellafield causes a build-up of artificial radionuclides in the marine environment and this is in clear violation of the principles of sustainable development. Whilst dumping of radioactive waste at sea is illegal, paradoxically, the discharge continuously via a pipeline is not illegal. Greenpeace believes that disposal of all radioactive waste into the sea should be banned including that discharged from land-based pipelines (Greenpeace 2000).

In parallel to its cessation target for hazardous chemicals, in 1998 OSPAR also adopted a strategy to tackle radioactive pollution within the same timeframe of 2020 (OSPAR 1998c). This strategy required progressive and substantial reduction in discharges, emissions and losses of radioactive substance to the marine environment, with the ultimate aim of achieving levels near background or close to zero. Unfortunately, implementation of the strategy has been fundamentally limited by ongoing (and over some periods, increasing) discharges from the reprocessing of spent nuclear fuels, an issue of long-standing disagreement between the Governments of Ireland, Denmark and Norway and the UK (OSPAR 2003). In the end, it seems inevitable that real progress will only be achieved as existing nuclear facilities reach the end of their working lives, rather than through any radical change in policy or practice. And in the meantime, the legacy of radioactive pollution of marine ecosystems in the North-East Atlantic will continue to grow.

4.3   Nutrient Pollution and Marine 'Dead Zones'

Excessive inputs of nutrients into the marine environment from man's activities, particularly nitrogen, has been linked with the formation of areas on the seabed in coastal areas which are depleted in oxygen (hypoxic) in different regions around the world (Diaz 2001). The lack of oxygen in such hypoxic zones on the seabed makes it impossible for many bottom-dwelling animals to survive and the areas have been called 'dead zones'. In some regions, dead zones extend across vast areas and research has indicated that the problem is growing. The number of dead zones has risen every decade since the 1970s with the total number being recently estimated as up to 200 (UNEP 2006c). Plate 4.1 shows a world map of dead zones.

Nutrient pollution, in the form of nitrogen or phosphorous, reaches coastal waters from a variety of human sources including run-off of fertilisers into rivers, sewage discharges and atmospheric pollution from the burning of fossil fuels (Rabalais et al. 2002). Excess nutrient pollution in coastal waters can cause increased numbers of phytoplankton (minute plants that float in the water) and zooplankton (tiny animals that float on the water) as well as marked changes in species composition. As the phytoplankton and zooplankton die and sink to the depths they are consumed by microbes on the seabed. Increases in the number of these microbes as a result of increased numbers of dead phytoplankton and zooplankton, causes oxygen to be used up in bottom waters (Dodds 2006; Ferber 2004). Fully oxygenated seawater contains dissolved oxygen at a level of about 10 parts per million (ppm) but, once oxygen levels in the water falls to 5 ppm, fish and other marine animals have trouble breathing (Raloff 2004). Hypoxic zones are defined as areas where the oxygen level has fallen below 2 ppm. While fish vacate areas when levels fall below 2 ppm, other less mobile sediment-dwelling animals cannot escape and begin to die at around 1.5 ppm oxygen (Raloff 2004; Bonsdorff

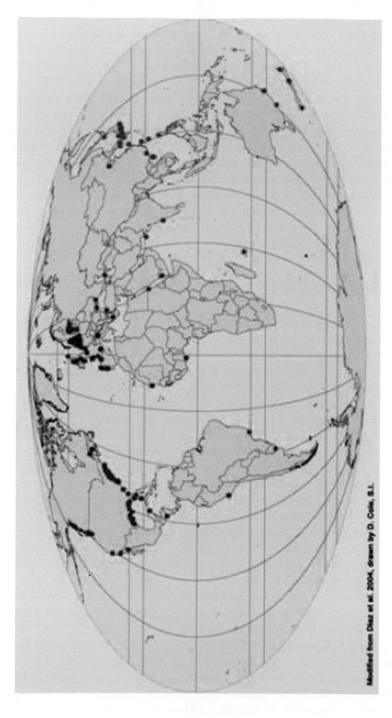

Plate 4.1 Global map of 199 coastal oxygen depletion zones related to anthropogenic eutrophication (UNEP and WHRC 2007)

Modified from Diaz et al. 2004, drawn by D. Cole, S.I.

et al. 2002). Biodiversity is thus diminished on the seabed as many animals cannot survive, even though higher in the water column there is still sufficient oxygen to support animal life.

The term 'dead zone' actually refers to the failure to capture fish, shrimp and crabs using bottom-dragging trawls (Rabalais et al. 2002). Dead zones in the Baltic and Black Sea led to the demise of demersal (bottom) fisheries in these areas (Diaz 2001).

The largest dead zones are found in coastal areas of the Baltic Sea (84,000 km²), northern Gulf of Mexico (21,000 km²) and, until recently, the northwestern shelf of the Black Sea (40,000 km²). Smaller and less frequently occurring areas of hypoxia occur in the northern Adriatic Sea, the southern bight of the North Sea and in many U.S. coastal and estuarine areas including New York Bight and Chesapeake Bay (Rabalais et al. 2002). Recent research shows that hypoxic areas are now also occurring off South America, China, Japan, southeast Australia and New Zealand. Some of the hypoxic zones are fleeting, whereas others persist for large proportions of the year (UNEP 2006c). For example, the dead zone in the northern Gulf of Mexico is dominant from spring through to late summer, but rare in the autumn and winter (Rabalais et al. 2002).

Severe bottom hypoxia as a result of nutrient pollution was first recorded around 1950 for the Baltic and Gulf of Mexico (Karlson et al. 2002; Rabalais et al. 2002). Accelerated growth of the hypoxia zone in the Gulf of Mexico follows the exponential growth of fertiliser use beginning in the 1950s (Rabalais et al. 2002). Elsewhere, other sources may also be responsible for nutrient pollution, such as municipal and industrial waste water and atmospheric deposition, and these have also increased since the 1950s (Rabalais et al. 2002). In the Baltic, there is clear evidence that the excess use of fertilisers is associated with hypoxic bottom water (Karlson et al. 2002).

The dead zone in the northwestern Black Sea in the 1970s and 1980s covered up to 40,000 km². Since then there has been some recovery, most likely due to the reduction in the use of agricultural fertilisers. This occurred as a result of economic collapse of the former Soviet Union and declines in subsidies for fertilisers. Less fertiliser input to the Danube River was accompanied by signs of recovery of both pelagic (open water) and benthic (bottom-dwelling) ecosystems of the Black Sea. By 1999, the hypoxic area had receded to less than 1,000 km². However, there has been no recovery of benthic seaweed beds and most fish stocks are still depleted (see Rabalais 2002).

Hypoxic zones are increasing in coastal regions around the world, causing reduction of biodiversity and in some cases impacts on fisheries. In the case of the Gulf of Mexico, Dodds (2006) therefore suggested that it would be prudent to reduce inputs of nitrogen and phosphorous to the coastal marine environment. This would seem applicable to all areas of the world where there is high nitrogen and phosphorous input to coastal waters. Greenpeace supports sustainable methods of farming and a move towards the cessation of the burning of fossil fuels, both of which would contribute to lessening nutrient inputs to the coastal marine environment.

4.4   Oil Pollution

Large oil spills to the marine environment can be catastrophic for wildlife, and
health impacts can be long-lasting if the oil pollution persists. While large spills
make the headlines because of their dramatic effects, smaller spills actually occur
every day. Sources of such spills include ship spills, offshore oil drilling operations and routine ship and car maintenance (Greenpeace 2006c). For example,
from 1990 to 1999, there were 513 spills from tankers and tank barges in US
coastal waters of at least 100 gallons (3791) in size (National Research Council
2003). In the North Sea, lawful discharges of oil from offshore oil and gas
installations accounted for the overwhelming bulk of oil inputs from this sector
(OSPAR 2004).

   While the size of a spill is important, the amount of damage done also depends
on other factors including the type of oil spilled, the location of the spill and
weather conditions (Greenpeace 2006c). Plate 4.2 shows contamination of a sandy
beach in Beirut following a major oil spill from the Jieh power plant in 2006.

Plate 4.2 View of oil slick which is covering Ramleh Baydah, the only public sandy beach
in Beirut, Lebanon. The oil spill occurred after Israeli bombers targeted a power plant in Jiyeh,
releasing 10,000-35,000 t of heavy fuel oil into the Mediterranean Sea (Greenpeace/Thomas
Kukovec).

4.4.1 Effects of Oil Spills

Oil spills can have devastating impacts on the environment, leaving dead or moribund animals and oil-coated shorelines. Seabirds and marine mammals are particularly affected, with feathers or fur becoming coated in oil. Oil-coated feathers of seabirds or the pelage of a marine mammal can destroy their waterproofing and insulating characteristics and this can lead to death from hypothermia. These animals may also be poisoned by the ingestion of oil as they try to clean themselves or if their prey is contaminated. In the long-term, there is evidence that continual exposure to low levels of oil can have a significant effect on the survival and reproductive performance of seabirds and some sea mammals (National Research Council 2003).

Chemicals in oil that are known to kill marine organisms include monocyclic aromatic hydrocarbons consisting mainly of benzene, toluene, ethylbenzene, and xylene (BTEX). These chemicals usually evaporate to the atmosphere from spilled oil within a few days so that acute fish kills are thought to be limited to the first 1 or 2 weeks after a spill. However, certain other chemicals (e.g. polycyclic aromatic hydrocarbons – PAHs) are much more long-lived and can be toxic in the long term (Short et al. 2003).

Two major oil spills and their impacts on marine wildlife are discussed here – one which took place in the far north in Alaskan waters (*Exxon Valdez*) and the other in more temperate waters of northern Spain (*Prestige*).

4.4.2 Exxon Valdez Oil Spill, Alaska 1989

The *Exxon Valdez* ran aground in March 1989 and spilled an estimated 42,000 t of crude oil into Prince William Sound on the margin of the northern Gulf of Alaska (Peterson 2001; Peterson et al. 2003). It resulted in the contamination of at least 1,990 km of pristine shoreline (Peterson et al. 2003). The spill was particularly damaging because it affected a coastal system which was notably rich in seabirds, marine mammals and shoreline-dependent species. Figure 4.2 shows an oiled sea otter at a rehabilitation centre.

The immediate effects of the oil spill on seabirds of all types was large (Peterson 2001) and it has been estimated that 250,000 birds were killed (National Research Council 2003). Longer-term effects were also found in many seabird species, including declines in abundance and changes in distribution. For example, research on several seabird species along oiled shorelines showed that there were persistent reductions in their abundance through 1998, which, in turn, may be related to reduced abundances of forage fish for them to prey on, at least in part due to the oil spill. A study of black oystercatchers (*Haematopus bachmani* Audubon) which fed on oiled mussel beds in 1991/2 showed that they produced fewer and smaller eggs on nesting than birds at non-oiled sites, an indication of chronic reproductive impairment due to oil contamination (reviewed by Peterson 2001).

Fig. 4.2 Oiled sea otter at rehabilitation centre in Valdez after oil spill (Greenpeace/Merjenburgh)

Marine mammals such as sea otters (*Enhydra lutris*) and harbour seals (*Phoca vitulina*) were affected by the oil spill because of their need to come to the surface to breathe, and their use of oiled shoreline rocks as haul-outs. It was estimated that up to 2,800 sea otters and at least 302 harbour seals were directly killed by the spill in Prince William Sound. Following initial impacts, both these species showed several years of delay in recovery in the spill area, indicating long-term effects of the spill (Peterson 2001). Research has shown that oil contamination from the spill was still evident on coastlines in Prince William Sound when tested 12 years after the spill (Short et al. 2004) and was also evident 10 years after the spill on sites in the Gulf of Alaska distant from the spill's origin (Irvine et al. 2006). A study in 2003 noted that oil contamination persisted in the lower inter-tidal zone of some beaches near to the spill, areas in which sea otters and ducks would be likely to encounter the oil while foraging (Short et al. 2006). A study on sea otters at Knights Island showed that they had not fully recovered in this area by 2000 due to higher rates of mortality and emigration from this oiled area (Bodkin et al. 2002).

In addition to marine mammals and birds, many other species were negatively affected by the oil spill. There was mass mortality of macroalgae (seaweed) and some invertebrates on oiled shores as a result of smothering, chemical toxicity and physical displacement from habitat by the use of pressurised water for clean-up. Research suggested that oil contamination was responsible for killing pink salmon (*Oncorhynchus gorbuscha* Walbaum) embryos in contaminated streams for at least 4 years after the spill (cited by Peterson et al. 2003). A recent study funded by Exxon Mobil concluded that subsurface oil contamination from the spill in 2004/5 was not having a detrimental impact on a particular species of fish, but the study came under heavy criticism from an independent scientist who has worked on impacts of the spill (Renner 2006).

4.4.3   Prestige Oil Spill, Spain 2002

On November 13, 2002, the oil tanker *Prestige*, loaded with a cargo of 77,000 t of heavy bunker oil, sank 130 miles off the coast of Spain (Garza-Gil et al. 2006). It is estimated that 63,000 t of oil has been released (Morales-Caselles et al. 2006), the rest remaining in the vessel's tanks on the seafloor. Almost all the coastline of Galicia (Northwest Spain) was affected by the spill, plus some points in Northern Spain and in Southwest France (Garza-Gil et al. 2006). A significant amount of spilled oil was recovered at sea (about 23,000 t by January 2003) and some was collected at the coast by volunteers, but this left over 50,000 t still at large in the marine environment or in the Prestige (García Pérez 2003). The impacts of the oil spill affected areas of ecological and touristic value and the problem was worsened by the lack of response to immediate needs in the early weeks of the crisis, allegedly due to weakness of decisional frameworks and organisational capacity (Albaigés 2006).

Over 23,000 oiled birds were collected after the spill (see, for example, Fig. 4.3), although the total number of birds affected was estimated to be between 115,000 and 230,000. The most affected seabirds were the common guillemot (*Uria aalgae*),

Fig. 4.3 A cormorant badly affected by crude oil, on a beach at Valcovo, Galicia, which spilled from the tanker Prestige (Greenpeace/Pedro Armestre)

followed by the razorbill (*Alca torda*), Atlantic puffin (*Fratecula artica*) and northern gannet (*Sula bassana*) (Zuberogoitia et al. 2006). Hundreds of resident European shags (*Phalacrocorax aristotelis*) also died (Martínez-Abraín et al. 2006). A study of European shags at their main breeding ground following the spill showed that the population was reduced and showed signs of reproductive impairment (Velando et al. 2005). For instance, the breeding success was 50% less in oiled colonies in 2003 compared to un-oiled ones, and chick condition was poorer. The study results suggested that reproductive impairment in the birds could be due to reduced availability of their prey, sandeels, which were heavily impacted by the spill. It is also possible that the birds could have been suffering from negative health effects of oil exposure. A study on peregrine falcons (*Falco peregrinus*) in the North of Spain, birds that can be affected by contaminated prey, concluded that the oil spill had caused increased rates of adult mortality and reduced fertility (Zuberogoitia et al. 2006).

A study on the numbers of species of invertebrates in the inter-tidal zone on affected beaches along the Galician coast found decreases in the number of polychaetes, insects and semi-terrestrial crustaceans. The most affected beaches lost up to 66.7% of the total species richness (total number of species) (de la Huz et al. 2005). A study on four bottom-dwelling marine species found significant reductions in the abundance of Norway lobster (*Nephrops norvegicus*), Pandalid shrimp (*Plesionika heterocarpus*) and the four-spot meagrin (*Lepidorhombus boscii*) in the Galician and Cantabrian Sea shelves following the oil spill (Sánchez et al. 2006). By 2004, there were noteworthy recoveries in the abundance of the Pandalid shrimp and four-spot meagrin. A study on the toxicity of oil-contaminated sediments from the spill investigated whether this level of contamination would have impacts on juvenile seabream fish (*Sparus aurata*) in a laboratory setting (Morales-Caselles et al. 2006). The results showed that increasing the sediment concentrations of chemicals from the oil pollution increased toxic effects in the fish. A study on mussels (*Mytilus galloprovincialis*) collected from the Bay of Biscay, Galicia, in 2003 indicated that they had been exposed to toxic chemicals and had disturbed health, results which provided preliminary evidence of the effects of the oil spill (Marigómez et al. 2006).

4.4.4 Solutions

In the past, Greenpeace campaigned for a ban on the use of single-hulled oil tankers so that only double-hulled vessels would be permitted. Regulations for a global phase out of single-hulled oil tankers by the International Maritime Organization were eventually brought into place in 2005.

To ensure effective clean-up of oil spills from tankers, Greenpeace is demanding full and unlimited liability through a chain of responsibilities, including the owners, managers and operators of a vessel and of any charterers or owners of the cargo, ensuring that the industry pays for the damage caused by accidents.

As discussed in Chapter 7, Greenpeace is recommending the establishment of a network of Marine Reserves (fully protected marine areas), covering at least 40%

of the world's oceans, in order to contribute to the overall protection, and in many cases recovery, of marine ecosystems. In areas where there has been damage caused by large oil spills, there can also be a need to establish Marine Reserves to allow the recovery of impacted marine ecosystems.

In the light of disasters and toxicity caused by oil spills, and the emerging threats of climate change, Greenpeace advocates the need to phase out the use of oil and move towards clean, renewable energy that can meet our needs without threatening the environment, now and into the future.

4.5   Plastic Debris

Solid materials, typically waste, that has found its way to the marine environment is called marine debris.

It is probably a common conception that marine debris consists of just a few pieces of rubbish scattered along the strand line of beaches and is of no harm to anyone. Unfortunately this is not the case. Marine debris has become a pervasive pollution problem affecting all of the world's oceans. It is known to be the cause of injuries and deaths of numerous marine animals and birds, either because they become entangled in it or they mistake it for prey and eat it. For a detailed review, see Allsopp et al. (2006).

Plastic and synthetic materials are the most common types of marine debris and cause the most problems for marine animals and birds. At least 267 different species are known to have suffered from entanglement or ingestion of marine debris, including seabirds, turtles, seals, sea lions, whales and fish.

The scale of contamination of the marine environment by plastic debris is vast. It is found floating in all the world's oceans, everywhere from polar regions to the equator. The seabed, especially near to coastal regions, is also contaminated – predominantly with plastic bags. Plastic is also ubiquitous on beaches everywhere, from populous regions to the shores of very remote uninhabited islands.

Attempts to address the problem of marine debris range from international legislation to prevent shipping from dumping plastic at sea, and campaigns to prevent losses due to poor industrial practice, to beach and seabed clean-up operations and public awareness campaigns. Plastic debris originates from a wide and diverse range of sources. Estimates suggest that much of what is found at sea originates on the land. The effect of coastal littering and dumping is compounded by vectors such as rivers and storm drains discharging litter from inland urban areas.

It is the very properties that make plastics so useful, their stability and resistance to degradation, which causes them to be so problematic after they have served their purpose. These materials persist in the environment and are not readily degraded or processed by natural biological mechanisms. However, plastics in the ocean are weathered; broken up either mechanically or by the action of sunlight into smaller and smaller fragments. Eventually, fragments are reduced to tiny pieces the size of grains of sand. These particles have been found suspended in seawater and on the seabed in sediments. Even such tiny particles may be causing harm to the marine environment since they have been shown to be ingested by small sea creatures and may concentrate persistent organic pollutants (POPs) present in the seas.

4.5.1 Sources of Marine Debris

It has been estimated that around 80% of marine debris is from land-based sources and the remaining 20% is from ocean-based sources. The sources can be categorised into four major groups:

- Tourism-related litter at the coast: this includes litter left by beach-goers such as food and beverage packaging, cigarettes and plastic beach toys.
- Sewage-related debris: this includes water from storm drains and combined sewer overflows which discharge waste water directly into the sea or rivers during heavy rainfall. These waste waters carry with them garbage such as street litter, condoms and syringes.
- Fishing-related debris: this includes fishing lines and nets, fishing pots and strapping bands from bait boxes that are lost accidentally by commercial fishing boats or are deliberately dumped into the ocean.
- Wastes from ships and boats: this includes garbage that is accidentally or deliberately dumped overboard.

Plate 4.3, taken in Lebanon, illustrates the severe extent to which plastic wastes can be found littering beaches in some regions. Plate 4.4 shows the extent of plastic pollution in parts of Manila Bay in 2006.

Plate 4.3 A young boy plays with syringes on a beach in Lebanon, surrounded by garbage and other debris that has been washed up by the tide (Greenpeace/Serji)

Plate 4.4 A local boatman swimming through rubbish in Manila Bay, Philippines, in 2006 (Greenpeace/Gavin Newman)

Huge volumes of non-organic wastes, including plastics and synthetics, are produced in more developed, industrialised countries. Conversely, in less developed and more rural economies, generally a much smaller amount of these non-biodegradable persistent wastes are produced. However, in the future, as less developed countries become more industrialised, it is likely that they will also produce more plastic and synthetic wastes and this will increase further the threat of pollution of the marine environment.

4.5.2 Harm to Marine Wildlife

Countless marine animals and seabirds become entangled in marine debris or ingest it. This can cause them serious harm and often results in their death.

4.5.2.1 Entanglement in Marine Debris

Marine debris which is known to cause entanglement includes derelict fishing gear such as nets and mono-filament line and also six-pack rings and fishing bait box strapping bands. This debris can cause death by drowning, suffocation, strangulation, starvation through reduced feeding efficiency, and injuries. Particularly affected are seals and sea lions, probably due to their very inquisitive nature of investigating objects in their environment. Entanglement rates in these animals of up to 7.9% of a population have been recorded. Furthermore, in some instances entanglement is a threat to the recovery of already reduced population sizes.

An estimated 58% of seal and sea lion species are known to have been affected by entanglement, including the Hawaiian monk seal (*Monachus schauinslandi*), Australian sea lions (*Neophoca cinerea*), New Zealand fur seals (*Arctocephalus fosteri*) and species in the Southern Ocean.

Whales, dolphins, porpoises, turtles, manatees and seabirds have all been reported to have suffered from entanglement. Many different species of whale and turtle have been reported to have been tangled in plastic (see, for example, Fig. 4.4). Manatees (*Trichechus manatus latirostris*) in Florida have been found with scars or missing flippers due to entanglement. Fifty-one species of seabirds are also known to have been affected.

Derelict fishing gear also causes damage to coral reefs when nets or lines get snagged by the reef and break it off. Finally, discarded or lost fishing nets (Fig. 4.5) and pots can continue to trap and catch fish even when they are no longer in use. This phenomenon is known as ghost fishing, and it can result in the capture of large quantities of marine organisms. Consequently, it has become a concern with regard to the conservation of fish stocks in some areas and has resulted in economic losses for fisheries.

4.5.2.2 Ingestion of Marine Debris

Ingestion of marine debris is known to particularly affect sea turtles and seabirds but is also a problem for marine mammals and fish. Ingestion is generally thought

Fig. 4.4 Sea turtle entangled in discarded fishing gear in the Mediterranean Sea (Greenpeace/ Marco Care). The turtle was subsequently freed by the crew of the Greenpeace ship Rainbow Warrior

Fig. 4.5 Discarded fishing gear
('ghost net') drifting in the Central
North Pacific Ocean (Greenpeace/
Alex Hofford)

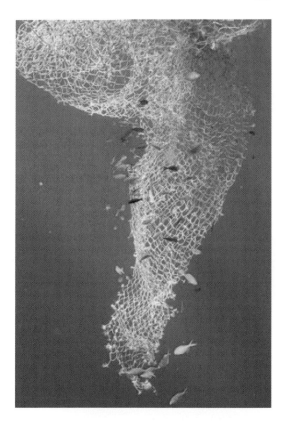

to occur because the marine debris is mistaken for prey. Most of that which is erroneously ingested is plastic. Different types of debris are ingested by marine animals, including plastic bags, plastic pellets and fragments of plastic that have been broken up from larger items.

The biggest threat from ingestion occurs when it blocks the digestive tract, or fills the stomach, resulting in malnutrition, starvation and potentially death. Studies have shown that a high proportion (about 50–80%) of sea turtles found dead are known to have ingested marine debris. This can have a negative impact on turtle populations. In young turtles, a major problem is dietary dilution in which debris takes up some of the gut capacity and threatens their ability to take on necessary quantities of food.

For seabirds, 111 out of 312 species are known to have ingested debris and it can affect a large percentage of a population (up to 80%). Moreover, plastic debris is also known to be passed to the chicks in regurgitated food from their parents. One harmful effect from plastic ingestion in birds is weight loss due, for example, to a falsely sated appetite and failure to put on adequate fat stores for migration and reproduction.

4.5.3   Potential Invasion of Alien Species

Plastic debris which floats on the oceans can act as rafts for small sea-creatures to grow and travel on. Plastic can travel for long distances and therefore there is a possibility that marine animals and plants may travel to areas where they are non-native. Plastic with different sorts of animals and plants have been found in the oceans in areas remote from their source. This represents a potential threat for the marine environment should an alien species become established. It is postulated that the slow speed at which plastic debris crosses oceans makes it an ideal vehicle for this; the organisms have plenty of time to adapt to different water and climatic conditions.

4.5.4   Marine Debris Around the World

Many studies have been carried out in different countries and oceans estimating the quantity of plastic on beaches, the sea floor, in the water column, and on the sea surface.

Most of these studies have focused, partly for reasons of practicality, on large (macro) debris. A limited body of literature also exists concerning small to microscopic particles (micro debris). The results show that marine debris is ubiquitous in the world's oceans and shorelines. Plate 4.5 shows plastic micro debris from a typical tow-net sample collected well off shore in the Atlantic Ocean. Higher quantities are found in the tropics and in the mid-latitudes compared to areas towards

Plate 4.5 Typical sample of plastic micro debris, collected from the surface of the Atlantic Ocean off the Azores, using a towed marine debris sampler, 2006 (Greenpeace/Gavin Newman)

the poles. It has been noted that high quantities are often found in shipping lanes, around fishing areas and in oceanic convergence zones.

- Floating marine debris: studies on different areas of the marine environment reported quantities of floating marine debris that were generally in the range of 0–1 items of debris per square kilometre. Higher values were reported in the English Channel ($10–100 +$ items/km^2) and Indonesia (more than four items in every square metre). Floating micro debris has been measured at much higher levels: the North Pacific Gyre, a debris convergence zone, was found to contain very high levels, that when extrapolated represent near to a million items per square kilometre.
- Seafloor debris: Research has shown that marine debris was present on the seafloor in several locations in European waters, and also in the USA, Caribbean and Indonesia. In European waters, the highest quantity recorded was 101,000 items/km^2 and in one location in Indonesia the equivalent of 690,000 items/km^2.
- Shoreline debris: Surveys of shorelines around the world have recorded the quantity of marine debris either as the number of items per km of shoreline or the number of items per square metre of shoreline. The highest values reported were for one location in Indonesia (up to 29.1 items/m) and in Sicily (up to 231 items/m).

4.5.5   Solutions

There are a number of global, international and national initiatives in place that are aimed at protecting the oceans from marine debris. The most far-reaching of these is the International Convention for the Prevention of Pollution from ships (MARPOL). Annex V of MARPOL was introduced in 1988 with the intention of banning the dumping of most garbage and all plastic materials from ships at sea. A total of 122 countries have ratified the treaty.

There is some evidence that the implementation of MARPOL has reduced the marine debris problem but other research shows that it does not appear to have had any positive impact. It must also be remembered that an estimated 80% of marine debris originates from sources on land. Even with total global compliance with MARPOL these sources would remain. Other measures to address marine debris include manual clean-up operations of shorelines and the sea floor as well as school and public education programmes.

While the above measures are important at preventing or reducing the problem of marine debris, the ultimate solution to waste prevention is to implement a responsible waste strategy, namely the concept of 'Zero Waste'. Such a strategy encompasses waste reduction, reuse and recycling as well as producer responsibility and ecodesign. Ultimately, this would mean reduction of the use of plastics and synthetics such that they are only used where absolutely necessary and where they have been designed for ease of recycling within existing recovery infrastructure. It is possible that biodegradable plastics could be used where plastic was deemed necessary but this could not be seen as an environmentally sound alternative unless they are known to break down rapidly to non-hazardous substances in natural environments.

Chapter 5
Increasing Greenhouse Gas Emissions: Impacts on the Marine Environment

Abstract Increasing carbon dioxide levels in the Earth's atmosphere, due largely to the combustion of fossil fuels and deforestation, is almost certainly causing changes in the Earth's climate. For the world's oceans, the effects of climate change include increases in sea temperature, increases in sea level and a reduction of sea-ice.

Sea Temperature: Research shows that the global ocean has warmed significantly since 1955. This is of concern because an increase in temperature could negatively affect the bodily (physiological) processes of marine organisms and their survival (see Section 5.2). In addition, many marine fish seek preferred temperatures and increasing temperature is therefore likely to affect their distribution. A further effect of increases in sea temperature is coral bleaching (see Section 5.2.2). This phenomenon has been increasingly observed on reefs worldwide since 1979 and, in some instances, has caused large-scale coral mortality.

Sea Level: Between 1961 and 2003, figures indicate that the global average sea level rose at an average rate of 1.8 (1.3–2.3) mm/year (see Section 5.3). Sea-level rise is caused by the addition of water to the oceans from the melting of land ice and the expansion of seawater due to warming. Sea-level rise may cause increased coastal erosion, more extensive coastal flooding, drowning of low-lying coastal areas and landward intrusion of seawater into estuaries and aquifers.

Sea Ice: The poles have already witnessed changes in response to climate change. In the Arctic there has been a reduction in the length of the sea-ice season and an increase in length of the melting season (see Section 5.5). Research shows that some species which depend on sea-ice appear to have been negatively affected by the changing conditions, including ringed seals (*Phoca hispida*) and polar bears (*Ursus maritimus*). In the West Antarctic peninsula there have been retreats of ice shelves, a rise in summer sea surface temperatures and a reduction of winter sea-ice in seas adjacent to the peninsula (see Section 5.6). The latter may be having a negative impact on Antarctic krill (*Euphausia superba*), a key species in the food web of the Southern Ocean.

Ocean Acidification: A direct and non-climate related impact of rising levels of carbon dioxide in the atmosphere is ocean acidification (see Section 5.7). As concentrations

M. Allsopp et al., *State of the World's Oceans,*
© Springer Science+Business Media B.V. 2009

increase, more carbon dioxide becomes dissolved into seawater and this has alrea
made the oceans slightly more acidic (by about 0.1 units pH). It is predicted that
process of increasing ocean acidification will continue in the future alongside th
impacts of climate change itself. This is a matter of great concern in relation to creatures
that build shells or have other calcified structures because these structures will start to
dissolve if seawater becomes too acidic.

Solutions: It is evident that climate change is already affecting the world's oceans
and marine life. Because of the detrimental impacts of climate change on the
oceans and on the terrestrial environment, action is urgently needed at source,
including a switch from fossil fuels to clean renewable energies and more effective
measures for energy conservation (see Section 5.8). In addition, the establishment
of large-scale marine reserves may lead to greater resilience of marine ecosystems
compared to exploited areas and this could help to combat predicted changes.

Keywords Greenhouse gases, CO_2, climate change, sea temperature, sea level,
Arctic, Antarctic, sea ice, acidification, coral, ocean circulation, seabirds, marine
mammals, krill, impacts.

5.1   Introduction

The Earth's atmosphere partly consists of a blanket of gases, known as greenhouse
gases, which trap enough heat to sustain life. For about a thousand years before the
Industrial Revolution began in the mid-18th century, the amount of greenhouse gases in
the atmosphere remained relatively constant. However, since the Industrial Revolution
the amount of greenhouse gases in the atmosphere has increased (IPCC 2001a). In
particular, carbon dioxide is the most important anthropogenic greenhouse gas and
its concentration in the atmosphere has increased primarily due to fossil fuel use and
deforestation. The global atmospheric concentration of carbon dioxide has increased
from a pre-industrial value of about 280 ppm to a level of 379 ppm in 2005. This
atmospheric level of carbon dioxide in 2005 exceeds by far the natural range over the
past 650,000 years (180–300 ppm) as determined from ice cores (IPCC 2007a).

Greenhouse gases, including carbon dioxide, absorb incoming solar energy and
outgoing (reflected) radiant energy (Roessig et al. 2004). An increase in carbon
dioxide since the Industrial Revolution has enhanced the absorption and retention
of radiation, and this has resulted in an increase in atmospheric temperatures with
impacts on the Earth's climate (IPCC 2001a; Roessig et al. 2004).

In their latest report, the Intergovernmental Panel on Climate Change (IPCC)
have stated that "*Warming of the climate is unequivocal, as is now evident from
observations of increases in global average air and ocean temperatures, widespread
melting of snow and ice, and rising global average sea level*" (IPCC 2007a).

Human-induced climate change is predicted to have profound impacts on the
world's oceans and their ecosystems if current activities continue. In some cases
there have already been effects due to climate change. Present and predicted effects
include increases in sea-surface temperature (see Section 5.2); increases in sea level

(see Section 5.3); and, decreases in sea-ice cover (see Sections 5.5 and 5.6). Changes in salinity and ocean circulation may also occur (see Section 5.4). In the northern hemisphere, the greatest changes in climate are predicted to occur in the Arctic (Weller 2004), while in the southern hemisphere, the greatest increases in atmospheric temperature are already occurring in the Western Antarctic Peninsula (Meredith and King 2005). Present and predicted impacts of climate change on Arctic and Antarctic ecosystems are discussed in Sections 5.5 and 5.6 respectively. Another impact of rising atmospheric levels of carbon dioxide includes an increase in the acidity of the ocean (see Section 5.7).

5.2 Sea Surface Temperature Increase

Research shows that the global ocean has warmed significantly since 1955 and that the ocean has absorbed more than 80% of the heat added to the climate system (IPCC 2007b). Observations since 1961 have shown that the average temperature of the global ocean has increased to depths of at least 3,000 m (IPCC 2007a). Over the period 1961–2003, the global average ocean temperature has risen by 0.10°C from the surface to a depth of 700 m (IPCC 2007c). Warming of the ocean causes seawater to expand, which, in turn, contributes to sea level rise (IPCC 2007a).

5.2.1 Ecological Impacts

Temperature can affect the physiological processes of living organisms. In the marine environment, many organisms live at temperatures that are close to their thermal (temperature) tolerances and any increases in sea temperature could have a negative impact on their physiological functioning and survival (Harley et al. 2006). Water temperatures can have a direct effect on spawning and the survival of juvenile fish (IPCC 2001b). Laboratory research has indicated that temperature affects the growth of some fish species, and modelling data for sockeye salmon (*Oncorhynchus nerka*) has suggested that elevated water temperature could decrease growth and increase mortality (see Roessig et al. 2004). It has been suggested that climate change could reduce the abundance of many marine species and increase the likelihood of local and, in some cases, global extinction (Harley et al. 2006). Increasing sea temperature is implicated as the primary cause of the many and widespread episodes of coral bleaching that have occurred on coral reefs worldwide in recent years (see Section 5.2.2 below).

Many marine fish seek preferred temperatures, such that increasing sea temperature is likely to impact on their distribution as well as their abundance. For example, the temperature of the western Mediterranean Sea has been rising for the last 20–30 years and there have been parallel increases in the abundance of some thermophilic species (which thrive at high temperatures) including algae, echinoderms and fish. Research has shown that there has been a warming-associated poleward shift for a Californian gastropod (*Kelletia kelletia*) and a Caribbean coral (see Harley et al. 2006). A northward shift in the distribution of some North Sea fish was found to occur in response to rising sea temperatures between 1977 and 2001 (Perry et al. 2005). In polar regions, fish have narrow temperature limits and even slight changes in temperature could cause a shift in their geographical distribution and affect their physiological performance (Roessig et al. 2004).

The wider implications of sea temperature rise may be complex and unpredictable. For example, recent warming trends in north-western Europe have led to earlier spawning of the mollusc *Macoma balthica*, but not to earlier spring phytoplankton blooms. This has caused a temporal mismatch between the mollusc larvae and their food supply. Furthermore, the larvae are now suffering from increased predation by shrimp whose peak abundance time has also shifted (Harley et al. 2006).

There is evidence that there have been changes in marine plankton abundance and community structure over recent decades in many areas of the world. Plankton (small plants) and zooplankton (small animals) lie at the base of the marine food web. All fish consume zooplankton in their larval stages and some adult fish depend on this food source (Hays et al. 2005). A study of plankton in the North Sea concluded that rising temperatures since the mid-1980s have modified the plankton community in a way that has reduced the survival of young cod (*Gadus morhua*) (Beaugrand et al. 2003). North Sea cod have suffered severe declines due to overfishing, and this further negative impact due to climate change may have exacerbated the problem and slowed recovery.

5.2.2 Coral Bleaching

Coral reefs exist in shallow tropical and subtropical seas within 30° north or south of the equator (Hoegh-Guldberg 2005). They are highly diverse ecosystems which support hundreds of thousands of species, many of which are not yet described by science (Hoegh-Guldberg 1999) (see Chapter 1, Section 1.4). Reefs act as breakwaters and protect inshore environments from wave action (Hoegh-Guldberg 2005). There is great concern that they are in worldwide decline due to man's activities, including overfishing and pollution. Since 1979, another threat has become increasingly apparent on reefs worldwide; that of widespread coral bleaching. This phenomenon occurs mainly in response to increases in sea temperature (Barton and Casey 2005). Coral bleaching, its consequences, and predictions of the extent of coral bleaching in the future due to sea-surface temperature increases from climate change, are all discussed below.

Reef-building corals have a symbiotic (mutually beneficial) relationship with certain species of algae which live within the coral and supply energy from photosynthesis. An increase in sea temperature can cause the partial or total loss of the symbiotic algae and/or the loss of algal pigments from the coral. The loss results in a paling in coral colour (Douglas 2003; Hoegh-Guldberg 1999). Since corals rapidly lose their colour and turn a brilliant white, the phenomenon has been named coral bleaching (Hoegh-Guldberg 1999). Plate 5.1 shows patches of bleached coral on the Great Barrier Reef, Australia.

Plate 5.1 Bleached coral on the Great Barrier Reef, Australia (Greenpeace/ Roger Grace)

Large-scale, or mass, coral bleaching often involves hundreds or even thousands of kilometres of coral reef being bleached almost simultaneously. Mass coral bleaching events began to be recorded from 1979 onwards. Hoegh-Guldberg (2005) noted that there have been six major global cycles of mass coral bleaching over the past 20 years, with a pattern of increasing frequency and intensity. For example, Fig. 5.1 shows the increase in frequency of coral bleaching events recorded for the Caribbean from 1983–1998. Since 1995, most coral reefs worldwide have been affected by mass coral bleaching. Bleaching is often temporary, with corals regaining colour once environmental conditions ameliorate but is deleterious to the corals and can be permanent, resulting in coral death.

Both field and laboratory studies have indicated that warmer than normal sea temperatures are the primary cause of mass coral bleaching (Hoegh-Guldberg 2005). Small increases of at least 1°C greater than the summer mean maximum temperature can lead to bleaching (Hoegh-Guldberg 1999, 2005). It has been suggested that global warming due to man's activities has most probably contributed to the mass coral bleaching episodes that have occurred throughout the world (Reaser et al. 2000). Elevated sea temperatures leading to mass coral bleaching are suggested to be the result of a steadily rising baseline of sea temperatures over the last 100 years, combined with regionally specific El Niño and La Niña events

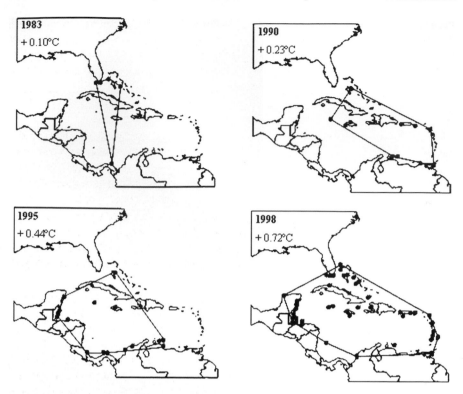

Fig. 5.1 Increase in frequency of coral bleaching events recorded for the Caribbean between 1983 and 1998 (Reproduced with permission from the Australian Institute of Marine Science and the Global Coral Reef Monitoring Network)

(Hoegh-Guldberg 1999; Reaser et al. 2000). The rise in sea temperature has increasingly brought corals close to the maximum temperature that they can tolerate. Mass coral bleaching events began to be observed after 1979 (Hoegh-Guldberg 1999). In 1998, there was a particularly severe impact in the Indian Ocean and Arabian region (see Table 5.1) as a consequence of increasing sea temperatures in the tropics and the 1997–1998 El-Niño-La Niña events (Australian Institute of Marine Science 2000). The authors noted that the losses in 1998 should be seen as temporary because many of the reefs should recover if further major bleaching events do not occur frequently.

Bleaching can have negative impacts on corals ranging from relatively mild, in the case of seasonal bleaching observed on many reefs, to large-scale mortality (Douglas 2003). The reproductive capacity and growth of corals can be reduced following bleaching (Hoegh-Guldberg 1999) and they may become more susceptible to disease (Reaser et al. 2000). Mortality close to 100% was observed in Indonesian and eastern Pacific reefs following a bleaching event in 1982–1983. Mortality of 46% of corals in the Western Indian Ocean was recorded after the

Table 5.1 Bleaching of coral reefs in various regions before and during 1998 (Adapted from Australian Institute of Marine Science 2000)

	Reef % destroyed pre-1998	Reef % destroyed in 1998
Arabian region	2	33
Wider Indian Ocean	13	46

1997–1998 bleaching event (Hoegh-Guldberg 1999, 2005). The extent of coral mortality appears to increase with the intensity of a bleaching event which, in turn, is determined by the size and length of time of the temperature increases above the maximum mean summer temperatures (Hoegh-Guldberg 1999). For instance, there was extensive coral bleaching on the Great Barrier Reef, Australia in 2001–2002 with significant coral mortality in the hottest patches but no damage in cooler areas (Hughes 2003). Other species that inhabit coral reefs may also be affected by coral bleaching. For instance, species dependent on coral for shelter and sustenance have shown little recovery from severe bleaching events, according to two studies cited by Roessig et al. (2004).

Records show that sea temperature is warming in response to climate change (IPCC 2007a). Many corals live very close to their upper temperature tolerance thresholds and the evidence on coral bleaching to date indicates that even small increases in sea temperature are likely to have large-scale impacts on the distribution of reef-building corals. In most areas of the tropics and subtropics, the thermal thresholds of corals could be exceeded by 2030–2050 (Hoegh-Guldberg 2005). Unless there is a change in the thermal tolerance of corals, it has been predicted that coral reef bleaching on a worldwide scale could become an annual or biannual event within 30–50 years (Donner et al. 2005). Hoegh-Guldeberg (1999) noted that it is hard to believe that coral reefs will be able to survive annual or even biannual bleaching events of the same scale as the severe bleaching event that occurred in 1998, and yet such sea surface temperatures are predicted by 2050. Even if corals are not killed outright by more persistent bleaching they may fail to reproduce.

It has been suggested that there are two possible ways in which corals could develop to cope with the predicted increase in sea surface temperatures. One is by acclimatisation, whereby their physiology is changed so they are more tolerant of higher temperatures, and the other is by adaptation wherein individuals within a population that are better able to cope with the new temperatures survive and increase by the process of natural selection (Hoegh-Guldberg 1999). However, there is no evidence that corals will be able to undergo the necessary changes to keep pace with the predicted temperature increases (Donner et al. 2005; Hoegh-Guldberg 2005; Jokiel and Brown 2004). Nevertheless, research on algal symbiotes (the algae that live with the coral) has found that unusual and more thermally tolerant algae have become more abundant on some reefs that have been severely affected by bleaching and mortality (Baker et al. 2004). It was suggested that this adaptive shift in the corals may increase the resistance of the recovering reefs to future bleaching events. Another response to increasing temperature may be changes in coral community structure, such that more thermally tolerant species become more dominant. For instance, corals of the genus *Porites* are more thermally tolerant than the branching forms of coral of the genera *Acopora* and *Pocillopora*. A decrease in coral reef habitat diversity could therefore result. Eventually, even the toughest species could be defeated by rising sea temperature. Consequently, it has been predicted that existing corals will become increasingly rare as sea temperature increases (Hoegh-Guldberg 2005).

It is estimated that 0.5–2 million species exist within coral reefs worldwide and many species are highly dependent on reefs for their existence. Species that rely on corals for food and shelter are likely to suffer if corals are removed by rising sea temperature (Hoegh-Guldberg 2005). The loss of coral reefs would not only impact biodiversity but also dependent coastal communities. It has been estimated that 100 million people rely on coral reefs for daily sustenance. In addition to fishing, coral reefs are also important for the tourism industry. Even from an entirely human perspective, it is therefore of great concern that both trends of the past century and predictive models suggest that coral bleaching events are likely to become more frequent and severe if the climate continues to warm and the frequency of high temperature episodes in the marine environment increases (Reaser et al. 2000). Donner et al. (2005) concluded that the global prognosis of coral bleaching is unlikely to change unless there is an accelerated effort to stabilise atmospheric greenhouse gas concentrations.

5.3 Sea-Level Rise

Sea-level rise can be caused by the expansion of water due to warming and by the addition of water to the oceans from the melting of land ice. In their most recent report, the IPCC noted that mountain glaciers and snow cover have declined globally and widespread decreases in glaciers and ice caps have contributed to sea-level rise. Research shows that losses from the ice sheets of Greenland and Antarctica are very likely to have contributed to sea-level rise over the period from 1993 to 2003 (IPCC 2007a). During that time, figures show that thermal expansion and the melting of land ice each account for about half of the observed sea-level rise, although there is some uncertainty in the estimates (IPCC 2007c).

Figure 5.2 shows graphically the rate of sea level rise over 10 years from 1992, derived from satellite data. According to the IPCC (2007a), between 1961 and 2003, the global average sea level rose at an average rate of 1.8 (1.3–2.3) mm/year. It is estimated that the total sea-level rise for the whole of the 20th century is 0.17 (0.12–0.22) m or 17 cm (IPCC 2007a).

Over the coming century, modelling predicts that the sea level will rise at an even greater rate than between 1961 and 2003. For instance, under one scenario proposed by the IPCC, by the mid-2090s, global sea level reaches 0.22–0.44 m (or 22–44 cm) above 1990 levels, and is rising at about 4 mm/year (IPCC 2007c). As a consequence of sea-level rise, extreme high water levels associated with storm waves and surges are predicted to increase in frequency (IPCC 2001b).

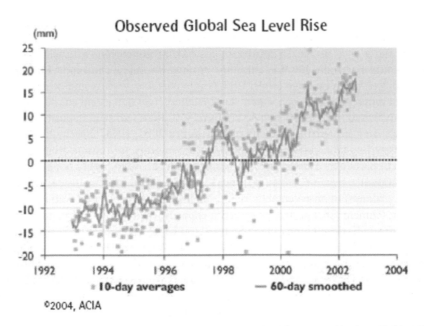

Fig. 5.2 Observed global sea-level rise. These data are derived from a satellite launched in 1992 (ACIA 2004)

In current estimations of sea-level increase, there is some uncertainty in predicting potential contributions to sea-level rise from the melting of the Greenland and Antarctic Ice Sheets, but the IPCC estimated that, between 1993 and 2003, melting of both ice sheets has contributed to sea-level rise (IPCC 2007d). Presently, there is a pattern of thinning of the West Antarctic Ice Sheet but thickening of the East Antarctic Ice Sheet (IPCC 2007d). For the Greenland Ice Sheet, the IPCC noted that thickening in central regions of Greenland has been more than offset by increased melting near the coast (IPCC 2007d). It was estimated that the melting of glaciers, together with water from the Antarctic and Greenland Ice Sheets, contributed to sea-level rise by approximately 0.7 ± 0.5 mm/year during 1961–2003 and 1.2 ± 0.4 mm/year during 1993–2003 (IPCC 2007d).

5.3.1 *Impacts of Sea-Level Rise on Society and Wildlife*

A rise in sea-level can result in increased coastal erosion, more extensive coastal flooding and drowning of low-lying coastal areas, higher storm-surge flooding and landward intrusion of seawater into estuaries and aquifers (Alley et al. 2005; Wong 2003; IPCC 2001b). Coastlines currently support a heavy concentration of human population and even small increases in sea-level rise could have substantial societal and economic impacts due to loss of habitat and saltwater intrusion into freshwater supplies (Alley et al. 2005). Estuarine and coastal marine wildlife would also be affected by loss of habitat.

One study cited by IPCC (2001b) suggested that, concurrent with sea-level rise, there would be tendencies for currently eroding shorelines to erode further and stable shorelines to begin to erode. Low-lying coastal regions would be exposed to potential inundation by ocean waters. For instance, river deltas and coastal wetlands (mangroves, salt marshes) which do not have sufficient sediment deposition, would be particularly susceptible to inundation (Hughes 2004; IPCC 2001b, 2007e). The IPCC identified river deltas as hotspots of vulnerability to impacts from sea-level rise and of particular societal concern because they are often highly populated areas (IPCC 2007e).

Many islands are threatened by rising sea levels and the United Nations Framework on Climate Change has produced a report in which it highlighted potential threats to small, developing, island States (UNFCCC 2005). The majority of these small island-States are located in the wider Caribbean and South Pacific regions. They share certain characteristics which underscore their overall vulnerability to sea-level rise. These include a concentration of population along the coastal zone, a high susceptibility to frequent and more intense tropical hurricanes and to associated storm surge and tsunamis, and dependence on water resources for freshwater supply that are highly sensitive to sea-level changes. For example, throughout the South Pacific region, frequent and more intense tropical cyclones, as well as climate-related and other extreme events, were experienced during the 1990s (UNFCCC 2005). In November 2005, the Papua New Guinea government started to move families from the Carteret atolls in the Pacific to a larger island, Bougainville, 60 miles away because the Carteret atolls are low-lying and are predicted to be uninhabitable by 2015 (Vidal 2005). If the sea level rises by 1m, the Maldives will disappear entirely (UNFCCC 2005).

Sea-level rise is predicted to cause losses in inter-tidal habitat and coastal wetlands. Wildlife which relies on these habitats is likely to be negatively impacted. For example, for a projected sea-level rise of 0.5 m on two islands in the Caribbean, which are known nesting sites for four species of marine turtles, it was estimated that up to 32% of the total current beach area could be lost. This would be likely to exacerbate the pressure already put on turtles by other human activities (Fish et al. 2005). A study which estimated sea-level rise in the Northwestern Hawaiian Islands by 2100 predicted that there would be significant loss of terrestrial habitat on two islands. It was proposed that this could impact negatively on the endangered Hawaiian monk seal (*Monachus schauinslandi*) and the threatened Hawaiian green sea turtle (*Chelonia mydas*) (Baker et al. 2006). A study on the predicted impacts of moderate sea-level rise on inter-tidal habitat at five sites in the USA estimated that there would be major habitat losses of 20–70% at four of the sites. Shorebirds depend on foraging habitat on tidal sand and mudflats and it was suggested that the predicted losses in the inter-tidal foraging zones at the four sites would be likely to result in major reductions in shorebird numbers (Galbraith et al. 2002).

Corals may be negatively affected by sea-level rise if the rise is rapid (Knowlton 2001). The algae living within the coral need light to photosynthesise and increasing water depth would reduce the available light. Thus, if the corals could not grow fast enough due to reduced light, they would effectively 'drown'. This would destroy the habitat upon which many fish and artisanal fishers depend (Roessig et al. 2004). More powerful or frequent storms could also cause increasing damage to coral reefs (Knowlton 2001).

5.4 Climate Change and Fishing

Climate variability is known to affect fish recruitment (the replenishment of stocks with juveniles) particularly towards the edges of species' ranges. However, according to Worm and Myers (2004), there is no evidence that worldwide stock declines are linked in any major way to climate change (though Antarctic krill may be an exception – see Section 5.7). On the other hand, there is abundant evidence that overfishing has resulted in significant declines in many fish species (see Chapter 1). Importantly, it has been suggested that heavily overfished stocks may be more sensitive to climate variability, due to a loss in biological diversity resulting in impaired resilience (Worm and Myers 2004). Therefore, it is possible for fishing pressure and climate change to act in concert and reduce exploited fish populations to below a population size from which they cannot easily recover (Harley et al. 2006). This re-emphasises the need to stop overfishing and a move to managing global fish stocks in a sustainable way (Worm and Myers 2004). In addition, even with accelerated action to switch to renewable energy systems, it will be not be possible to halt all of the impacts of climate change and, therefore, it has been suggested that further strategies are needed to protect marine ecosystems, particularly the establishment of marine reserves (Harley et al. 2006) (see Chapter 7).

5.5 The Ocean Conveyor Belt

The world's oceans affect climate in a multitude of ways. Solar radiation warms the surface waters by which heat is stored in the oceans and is redistributed by a global circulatory system (Rahmstorf 2002). Water is circulated slowly around different parts of the globe by the thermohaline circulation, so called because it is driven in part by temperature differences and in part by salinity differences. It is also known as the Ocean Conveyor Belt or global Meridional Overturning Circulation (MOC) (see Plate 5.2). It is responsible for transporting a huge amount of tropical heat to the northern Atlantic (via the Gulf Stream). During the wintertime, this heat is released to eastward-moving air masses, which has the effect of warming the climate over northern Europe (Broecker 1997). Climate change could potentially affect thermohaline circulation via surface warming of the oceans and via increased freshwater input from melting glaciers and sea-ice (Rahmstorf 2000). Climate models predict that this could weaken or even switch off the conveyor belt in the North

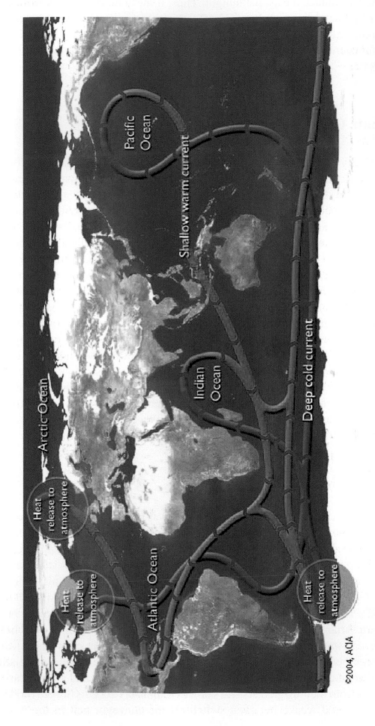

Plate 5.2 Global ocean circulation. Changes in global ocean circulation can lead to abrupt climate change. Such change can be initiated by increases in arctic precipitation and river runoff, and the melting of arctic snow and ice, because these lead to reduced salinity of ocean waters in the North Atlantic (ACIA 2004)

Atlantic and consequently there would be less of a warming effect over northern Europe (IPCC 2001a). The likelihood of any changes in thermohaline circulation is, however, not yet well established, but the possibility for an abrupt change and impact on climate is real (Weller 2004). In their most recent report, the IPCC (2007c) suggested that there is presently insufficient evidence to determine whether trends exist in MOC of the global ocean.

5.6 Climate Change and the Arctic

Over the last few decades there has been an overall trend of a substantial warming of the climate in the Arctic (Weller 2004). Average arctic temperatures increased at almost twice the global average rate in the past 100 years (IPCC 2007a). Research conducted from submarines has indicated that there has been about a 40% reduction in sea-ice thickness between 1958 and the 1990s (IPCC 2001a, 2007a). On the basis of submarine data and other research data, the IPCC noted that it is very likely that the average sea-ice thickness in the central Arctic has decreased by up to 1 m since the late 1980s (IPCC 2007d).

There has been a 10–15% decrease in the extent of the sea-ice in the spring and summer since the 1950s. Satellite data have indicated that there was a reduction in the length of the sea-ice season and an increase in the length of the Arctic melting season between 1979 and 1998 (IPCC 2001a). In their most recent report, the IPCC noted that satellite data since 1978 showed a reduction in the annual average arctic sea-ice extent by 2.7% (2.1–3.3) per decade, with larger decreases in summer of 7.4% (5–9.8) per decade (IPCC 2007a). Plate 5.3 shows satellite images of arctic sea-ice in 1979 and in 2003 clearly indicating a reduction in the extent of the ice.

Changes in the climate and sea-ice have not only been recorded by scientists but have also been witnessed by many indigenous peoples of the Arctic (Gibson and Schullinger 1998; Huntington and Fox 2004). Presently, the climate and sea-ice changes witnessed by indigenous peoples are increasingly beyond the range of their experience of the past (Weller 2004).

The Arctic Climate Impact Assessment (ACIA) is an international project on climate change co-ordinated by the Arctic Council (an inter-governmental forum) and the International Arctic Science Committee. It has been proposed by ACIA that predicted emissions of greenhouse gases are likely to result in further changes in the Arctic climate. The mean annual surface temperature north of 60° N is predicted to be 2–4°C higher by 2050 and 4–7°C higher toward the end of the 21st century. Sea-surface temperature is expected to increase and sea-ice extent and thickness are predicted to decrease (Weller 2004). The Arctic Ocean is predicted to be predominantly ice-free in summer by the end of the century (Smetacek and Nicol 2005). If the loss of sea-ice is as great as predicted by the Arctic Climate Impact Assessment models, then negative consequences are very likely within the next few decades for Arctic animals that depend on sea-ice for breeding and foraging (Loeng 2004). Many organisms are likely to be affected, including fish, birds, seals, whales and polar bears.

Observed sea ice September 1979

Observed sea ice September 2003

Source:
Arctic Climate Impact Assessment (ACIA), 2004.
Impacts of a Warming Arctic.

Plate 5.3 Observed sea-ice in September 1979 and September 2003 (ACIA 2004)

The ACIA predicted that climate warming will almost certainly cause a northward migration of arctic species towards cooler areas. Some species such as polar bears and seals, that inhabit the ice margins, will be at risk of extinction. Species which are presently abundant will be restricted in their range, which could have severe impacts on commercial fisheries, indigenous hunting and ecosystem function. The inherent loss in biodiversity of arctic ecosystems will probably also result in increased susceptibility to disease, insect pest infestations and parasites (Weller 2004).

5.6.1 Fish

Changes in sea temperature may directly affect fish by causing changes in their metabolic, growth and reproductive processes. There may also be changes in the growth and survival of smaller organisms which are their prey. The ACIA predicted that the distribution of Arctic fish will probably change, since most fish are sensitive to changes in water temperature. Some species, including Atlantic and Pacific cod (*Gadus morhua* and *G. macrocephalus*), Atlantic and Pacific herring (*Clupea harengus* and *C. pallasi*), walleye pollock (*Theragra chalcogramma*) and some flatfish are likely to move northward and may increase in abundance. Conversely, some other species, including capelin (*Mallotus villosus*), polar cod (*Boreogadus saida*) and Greenland halibut (*Reinhardtius hippoglossoides*) are likely to have a restricted range and decline in abundance (Loeng 2004).

5.6.2 Birds

The ACIA reported that seabirds are likely to be affected mostly by changes in their prey. For instance, possible changes in the distribution of capelin (*Mallotus villosus*) stocks in the Barents Sea could have major impacts on feeding for many Arctic seabirds in the region. The most sensitive seabird species to climate change are potentially those that have narrow food or habitat requirements (Loeng 2004). There is already concern regarding the ivory gull (*Pagophila eburnean*), a species that is closely associated with sea-ice. Research suggested there has been an 80% decline in the numbers of nesting ivory gulls and that therefore the population has declined. It was speculated that the cause of the decline is changes in sea-ice, causing an altered wintering habitat for the gulls (Gilchrist and Mallory 2005).

At Coats Island in northern Hudson Bay, Canadian Arctic, there were increased rates of egg loss and adult mortality of Brünnich's guillemot (*Uria lomvia*) in the late 1990s. This has been linked to increased mosquito numbers. It was suggested that higher temperatures had caused more favourable conditions for mosquitoes and that this had caused adverse impacts on the guillemots (Gaston et al. 2002).

5.6.3 Seals

Ice-living seals depend on sea-ice as pupping, moulting and resting platforms. Some species subsist on ice-associated prey species. The sea-ice must be sufficiently stable in order to rear pups. For example, in the Gulf of St. Lawrence, years with little or no sea-ice have resulted in almost no production of pups compared to hundreds of thousands in good sea-ice years. If spring sea-ice conditions follow current and projected trends, this will have dire consequences for the harp seal (*Phoca groenlandica*, Fig. 5.3) and hooded seals (*Cystophora cristata*) in the

Fig. 5.3 Harp seal (*Phoca groenlandica*) on sea ice in Canada (Greenpeace/Pierre Gleizes)

region. The ACIA predicted from modeling and/or from extrapolation of observed trends in climate change that other seal species dependent on sea-ice are also likely to decline, including ringed seals (*Phoca hispida*), spotted seals (*Phoca largha*) ribbon seals (*Phoca fasciata*) and possibly bearded seals (*Erignathus barbatus*). Conversely, more temperate seal species, the harbour seal (*Phoca vitulina*) and grey seal (*Halichoerus grypus*), are likely to expand their distribution in the Arctic if there is less sea-ice (Loeng 2004).

In Hudson Bay in the Canadian Arctic, lower snowfall, lower snow depth and warmer temperatures were recorded between 1990 and 2001 compared to previous records. Research on ringed seals (*Phoca hispida*) indicated lower pup production and pup survival, which appeared to be linked with lower snowfall. It was proposed that less snow meant less cover and protection for seal pups in their lairs, which in turn could have resulted in a lower survival rate. Higher temperatures may have also had an effect by melting the snow over birth lairs (Ferguson et al. 2005).

5.6.4   Polar Bears

Polar bears (*Ursus maritimus*) have inhabited virtually all the available sea-ice habitats throughout the circumpolar Arctic. Female polar bears from most populations use land areas to build dens to rear young but, otherwise, polar bears are heavily dependent on sea-ice as their habitat and feeding ground (Derocher et al. 2004). Sea-ice acts as a hunting platform for polar bears to feed on ice-associated seals (Fig. 5.4). In addition, although polar bears can swim if necessary, mothers and cubs rely on ice corridors to move from denning areas on land to hunting areas

Fig. 5.4 Polar bear on iceflow, Herald Island, Chuckchi Sea, Alaska (Greenpeace/Daniel Beltrá)

on sea-ice (Loeng 2004). Given predicted climate change and concurrent sea-ice retreat, the future of the polar bear is considered to be tenuous (Derocher et al. 2004). The ACIA proposed that there is likely to be a decline in polar bears and that they are at risk from becoming extinct (Weller 2004).

The first changes which may be predicted in a warming Arctic climate are the earlier break-up of the annual ice in spring and later time of freezing in autumn. This would result in a shorter period for polar bears to feed on the sea-ice and a longer fasting period off the ice. A shorter feeding period would result in reduced fat stores in the bears. Female bears with lower fat stores are likely to produce fewer cubs and have smaller cubs with lower survival rates (Derocher et al. 2004). In Hudson Bay, break-up of the sea-ice is now occurring about 2.5 weeks earlier than it did 30 years ago. Since the early 1980s, polar bears have been coming ashore in a poorer condition, birth rates have declined and it has been suggested that this is due to the earlier break-up of sea-ice (Loeng 2004). At Churchill in Manitoba, Western Hudson Bay, the snow has been returning later and later after the summer months and polar bears have needed to be tranquillised and airlifted north for them to access their natural habitat (Greenpeace 2004a).

Another way in which polar bears might be adversely affected by climate change is through reduced availability of their main prey – ringed seals (*Phoca hispida*). It is possible that ringed seals will decline as sea-ice is reduced and their habitat disappears. In addition, the thermal properties of polar bear dens may also be affected by warming temperatures, which could have a detrimental effect on cub survival (Derocher et al. 2004).

5.7   Climate Change and the Antarctic

Air temperature records for the Western Antarctic Peninsula show there has been a rapid rise in atmospheric temperature of nearly 3°C since 1951. This represents the greatest shift in climate in the Southern Hemisphere in recent decades. Concurrent with this change has been a rise in summer surface temperatures of the ocean adjacent to the West Antarctic Peninsula by 1°C (Meredith and King 2005). This will have contributed to a reduction in the extent of winter sea-ice which has been observed in recent years (Meredith and King 2005; Smetacek and Nicol 2005). At South Orkney, there has been an overall decreasing trend in the duration of sea-ice recorded over the past 90 years (Smith et al. 1999).

On the West Antarctic Peninsula, increasing atmospheric temperature has led to retreats in five ice shelves over the last century. This progressive retreat of the ice shelves led to the collapse of the Prince Gustav and parts of the Larsen ice shelves in 1995 (IPCC 2001a), and the dramatic break-up of the Larsen B Ice Shelf in 2002 (IPCC 2007d). The majority of glaciers in the region have retreated over the past 50 years and the average retreat rates are accelerating (Meredith and King 2005).

Changes in sea temperature, reduction of winter sea-ice and increased melting of glaciers and subsequent changes in adjacent ocean salinity are all factors which could affect, or are already affecting, life in the Antarctic (see below).

5.7.1   Impact of Climate Change on Krill

Antarctic krill (*Euphausia superba*) is a key species in the Southern Ocean food web. However, climate change in the region appears to be having a negative impact on their numbers. Since the mid-1980s, significantly lower population sizes of krill have been observed in the Antarctic Peninsula region (Loeb et al. 1997). One recent study investigated krill abundance in the productive southwest Atlantic sector of the Southern Ocean, where greater than 50% of the Southern Ocean krill stocks are located. It reported that, krill densities here had decreased significantly between 1976 and 2003 (see Fig. 5.5) (Atkinson et al. 2004). The study showed a direct temporal link between annual krill density and sea-ice cover. Summer krill density correlated with the extent of sea-ice the previous winter such that less ice the previous winter meant less krill. The sea-ice provides winter food from ice algae and is needed for the survival and growth of krill larvae.

Krill have a lifespan of about 6 years, and only a few 'good years' of krill breeding per decade are necessary to maintain good krill numbers. However, where ice conditions conducive to good breeding years for krill are deteriorating due to climate warming, this puts additional stresses on populations as the 6-year krill lifespan may fail to bridge the gap between 'good' years. Such a 'senescence' period is thought to have occurred in the western Antarctic region between 1980 and 1986 and it has been is hypothesised that this led to lower krill abundance in the early 1990s (Fraser and Hofmann 2003).

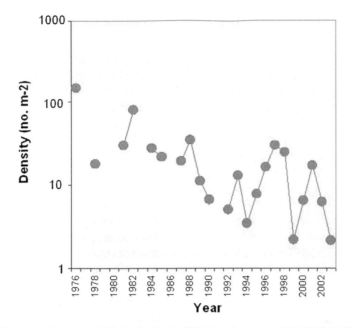

Fig. 5.5 Temporal changes in krill density in the SW Atlantic sector, 1976–2003 (Atkinson et al. 2004, reprinted by permission from Macmillan: Nature copyright 2004)

In addition to their reliance on sea-ice, Antarctic krill are also dependent on summer phytoplankton blooms as a food source (Atkinson et al. 2004). However, a study which recorded plankton community structure between 1990 and 1996 at Palmer Station, Antarctica, revealed a shift in the organisms making up the plankton from diatoms to cryptophytes (Moline et al. 2004). The change was linked to increased glacial melt water run-off that reduced the surface-water salinity. Cryptophytes are smaller than diatoms and are not grazed efficiently by Antarctic krill. It was suggested that the increase in cryptophytes could cause changes in the distribution of krill. Increased abundances of cryptophytes are also likely to favour the presence of salps (gelatinous marine animals, mainly *Salpa thompsoni*), which can graze cryptophytes. Salps can tolerate warmer water conditions than krill and their abundance in parts of the Southern Ocean appears to have increased while krill have decreased (Atkinson et al. 2004). Unlike krill, which support other species such as penguins, seals and whales, salps are not a preferred food for these creatures. If climate warming continues, the proportion of cryptophyte biomass to phytoplankton biomass would be expected to increase and would negatively impact on coastal food webs in the Antarctic (Moline et al. 2004).

Another factor which may influence krill numbers is sea temperature, as krill are believed to favour cold water. Summer sea temperatures in the Western Antarctic Peninsula were reported to have increased by 1°C over the past 50 years. It was suggested that temperature changes will have occurred in one of the key spawning and nursery areas for krill, which could have a negative impact on their population (Meredith and King 2005).

5.7.2   Effects of Declining Krill Abundance on Other Animals

As discussed above, it has been suggested that Antarctic krill (*Euphausia superba*) have undergone declines in abundance in recent years. Changes in this key species would be expected to have profound implications for the Southern Ocean food web. In particular, penguins, albatrosses, seals and whales are prone to krill shortages (Atkinson et al. 2004).

Research shows that penguins may already be affected by climate change and decreased krill abundance. Archaeological data show that Adélie penguins (*Pygoscelis adeliae*) have been permanent occupants of the Western Antarctic Peninsula for at least the past 600 years. On the other hand, chinstrap penguins (*Pygoscelis antarctica*) and gentoo penguins (*Pygoscelis papua*) appear to have expanded southward along the Western Antarctic Peninsula during the past 50 years, which could be linked to regional warming. The Adélie penguin is associated with winter pack ice whereas the chinstrap penguin occurs mainly in open water. Adélie penguin populations have been stable or declining slowly in the Western Antarctic Peninsula and it has been suggested that this may be due to a reduction in winter sea-ice (Smith et al. 1999). A recent study at the South Orkney Islands reported that the numbers of Adélie and chinstrap penguins have declined in the last 26 years, whereas gentoo penguins increased significantly (Forcada et al. 2006). These trends have occurred in parallel with regional long-term warming and significant reduction in the extent of sea-ice. Another study at the South Orkney Islands documented a reduction in the reproductive performance of Adélie and chinstrap penguins in 2000 (Lynnes et al. 2004). Results suggested that this may have been due to the abundance of krill falling below a critical level needed to support the normal breeding success of the penguins.

Fraser and Hofmann (2003) hypothesised that lower numbers of krill in the early 1990s resulted from a 'senescence' event in the 1980s and that this may have caused knock-on effects on Antarctic wildlife. For example, it was suggested that lower krill abundance may have been responsible for the decreasing populations of Adélie and chinstrap penguins observed since 1990 in some regions. In addition, decreasing trends in the birth weights of Antarctic fur seals and macaroni penguins in the early 1990s were reported, when the contribution (by weight) of krill in the diets of macaroni penguins began to decline significantly.

5.7.3   Effects of Changing Sea Temperature on Certain Antarctic Animals

Many benthic (bottom dwelling) Antarctic marine species are more sensitive to temperature variation than marine groups elsewhere. One study investigated the impact of water temperature increase on the biological functioning and survival of an Antarctic bivalve mollusc (*Laternula elliptica*), a limpet (*Nacella concinna*) and a scallop (*Adamussium colbecki*) (Peck et al. 2004). It was found that only a 1°C rise in summer sea temperatures negatively impacted their biological functions. For example, the Antarctic scallop lost the ability to swim with a 1°C temperature rise. It is therefore of great concern that most models predict a global sea temperature increase by 2°C or more in the next 100 years. A 2°C rise in sea temperature could cause population or species removal from the Southern Ocean of the species investigated in this study. It was suggested that most of Antarctica's 4,000 or more marine benthic species described to date would be at risk of population losses from a 1°C to 2°C increase in summer sea temperatures.

5.8 Ocean Acidification and Its Impacts on Marine Organisms

As stressed above, a large proportion of the carbon dioxide emitted to the atmosphere is absorbed into the world's oceans. The combination of carbon dioxide with water results in the formation of a weak acid called carbonic acid. The increasing carbon dioxide content of the atmosphere since industrialisation is therefore predicted to make the oceans more acidic. Over the past 200 years, since pre-industrial times, the oceans have absorbed about half of the carbon dioxide emissions produced from the burning of fossil fuels and deforestation. This has resulted in a lowering of the pH of the world ocean by about 0.1 units. Figure 5.6 shows graphically an increasing level of carbon dioxide in ocean waters at three different locations and a corresponding decrease in pH. Although a pH change of about 0.1 units may seem small, the change actually represents a considerable acidification of the oceans (Royal Society 2005). Moreover, the predicted rise in atmospheric carbon dioxide over the 21st century is estimated to give reductions in the average global ocean-surface pH of between 0.14 and 0.35 units (IPCC 2007a). Such a change in pH could have a detrimental impact on many marine organisms that build shells and other structures out of calcium carbonate (calcified organisms).

Calcified marine organisms include corals, certain crustaceans, echinoderms (e.g. sea urchins, starfish, brittle stars), certain molluscs and certain plankton and zooplankton. These organisms rely on the process of calcification to build their body/shell structures and could therefore be adversely affected by ocean acidification (Royal Society 2005). The threats of ocean acidification to these organisms are twofold. Firstly, their shells may start to disintegrate because calcium carbonate dissolves under acidic conditions (Doney 2006). Secondly, it is expected that calcifying organisms will find it more difficult to produce and maintain their shells and hard structures (Royal Society 2005). This is because the concentration of carbonate ions in seawater, which are needed in the building of calcified structures, will decrease as more carbon dioxide is absorbed by the oceans and the pH balance shifts (Doney 2006).

Calcified marine organisms use calcium carbonate in either of two different mineral forms – calcite or aragonite. Some calcite-secreting organisms also incorporate magnesium in calcified structures. Both aragonite and magnesium calcite are more

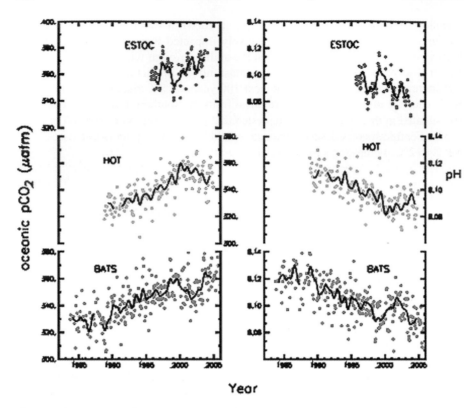

Fig. 5.6 Changes in surface oceanic carbon dioxide level and pH. Changes in surface oceanic pCO$_2$ (left; in µatm) and pH (right) from three time-series stations: European station for time-series in the ocean (ESTOC, 29°N, 15°W), Hawaii ocean time-series (HOT, 23°N, 158°W), Bermuda Atlantic time-series study (BATS, 31/32°N, 64°W) (IPCC 2007c)

soluble than normal calcite, which means that organisms using these materials may be especially susceptible to the effects of ocean acidification (Doney 2006). Shallow warm surface waters of the ocean are described as super-saturated with respect to aragonite and calcite, which means that these minerals have no tendency to dissolve. Colder waters are naturally less supersaturated than warm ones which makes the process of calcification by organisms more difficult. Calcified marine organisms that inhabit colder waters may therefore be the first to suffer from ocean acidification (Doney 2006). The Southern Ocean has been identified as being of particular concern (Royal Society 2005). However, even tropical waters may be affected. For instance, it has been estimated that the degree of saturation of aragonite in tropical waters has already decreased over the past century and, consequently, calcification has probably also decreased on some coral reefs (Kleypas et al. 1999).

It has been predicted that the surface waters of the Southern Ocean will begin to become under-saturated with respect to aragonite by the year 2050 and, by the year 2100, this under-saturation could extend throughout the entire Southern Ocean and into the subarctic Pacific Ocean. An experiment was conducted to determine whether a species of pteropod (swimming mollusc, *Clio pyramidata*) could survive under the seawater conditions predicted for 2100 (Orr et al. 2005). The results showed that the shell started to dissolve along the growing edge after just 48 hours. It was suggested that, if this organism could not grow its protective shell, it would not be expected to survive. Pteropods are an important food source for marine predators; for instance, they contribute to the diet of many species including zoo-plankton, North Pacific salmon (*Oncorhynchus* spp.), mackerel, herring, cod and even baleen whales (see Orr et al. 2005) The elimination of pteropods from the Southern Ocean would have unknown and potentially large-scale ramifications for the ecosystems. For instance, if predators switched to different prey, this could result in greater predation pressure on juvenile fishes (Royal Society 2005).

Photosynthetic corals need aragonite calcified structures in order to survive and it has been suggested that acidification of the oceans would result in significant declines in corals of the tropics and subtropics (Royal Society 2005). Orr et al. (2005) noted that there is substantial evidence that calcification rates will decrease in low latitude corals. For example, one experiment on corals common to Kaneohe Bay, Hawaii, simulated water carbon dioxide conditions predicted in the mid-to-late 21st century, and found that there was a large decrease in coral calcification in response to the elevated carbon dioxide (Langdon and Atkinson 2004). The effects of reduced calcification include weaker skeletons, increased susceptibility to erosion and reduced coral cover (Kleypas et al. 1999; Royal Society 2005). Coral reefs provide habitat and nursery grounds for numerous species (see Chapter 1, Section 1.4) and ocean acidification is likely to have major ramifications on the biodiversity and function of these ecosystems and have implications for ecosystems associated with reefs (e.g. seagrass, mangrove). They also support millions of people for subsistence food gathering and through tourism. In addition to tropical and subtropical coral reefs, it has also been predicted that the existence of cold water coral reefs are threatened by ocean acidification (Royal Society 2005).

Other species that are threatened by ocean acidification are some organisms in the Arctic and Antarctic that secrete magnesium calcite, which can be more soluble than aragonite. These include gorgonians, coralline red algae and echinoderms (see Orr et al. 2005). Both echinoderms and crustaceans in general have been identified as groups that may be particularly vulnerable to ocean acidification (Royal Society 2005). Ocean acidification may also affect other marine organisms in addition to calcifying organisms. The respiratory processes of fish and invertebrates may be affected by increased carbon dioxide, and body tissues may become acidified. Other effects could include decreased reproductive potential, slower growth or increased susceptibility to disease, although experimental evidence is currently lacking (Royal Society 2005).

A report on ocean acidification by The Royal Society, UK, concluded that there was no realistic way to reverse the widespread chemical effects of ocean acidification or the subsequent biological effects. It was suggested that the only viable and practical solution to minimise the long-term consequences of ocean acidification is to reduce carbon dioxide emissions into the atmosphere. Without significant action to do this there could be no place in the future oceans for many of the species and ecosystems we know today. It was concluded that ocean acidification is a powerful reason, in addition to climate change, for reducing global carbon dioxide emissions (Royal Society 2005).

5.9 The Way Forward

The above discussion demonstrates that marine organisms in polar regions may already be suffering from the impacts of climate change. Furthermore, future predictions suggest that marine ecosystems in polar regions could be severely impacted by climate change, and marine ecosystems elsewhere could also be negatively affected. In addition, increasing carbon dioxide levels from the burning of fossil fuels could cause further acidification of the oceans and threaten the existence of marine calcifying organisms.

Because of the threats of climate change to marine and terrestrial ecosystems, Greenpeace is campaigning for a switch from fossil fuels to using renewable energy, such as solar and wind power, which does not emit carbon dioxide. Governments and corporations need to shift away from polluting technologies to clean renewable energies with the aim of providing sustainable energy for all. With regard to the marine environment, the additional measure of designating large-scale marine reserves is advocated by Greenpeace because this may help to mitigate the potentially harmful effects of climate change, and is vital for many other reasons (see Chapter 7). Marine reserves encourage stable populations of marine organisms and intact communities which appear to be more resilient to climatic disturbances. Consequently marine reserves may help minimise the risk of population collapses of marine organisms, community disruption and biodiversity loss (Harley et al. 2006). A well-designed global network of reserves could act as a series of stepping stones, providing refugia for populations whose distribution is moving with climate change (Marshall et al. 2005).

Chapter 6
Equity

Abstract Fundamental changes need to be made to the way in which our oceans are managed. Presently, overfishing in many regions has depleted fish stocks. Furthermore, trading agreements can be unfair to developing countries. To bring about solutions, Greenpeace is working to promote the measures that are necessary to achieving sustainable and equitable management of our oceans. A number of the important issues relating to such sustainability and equitable management that have been part of Greenpeace campaigns are discussed in this chapter.

Illegal, Unreported and Unregulated (IUU) Fishing: IUU fishing, also known as pirate fishing, is a prevalent and widespread problem. Pirate fishers do not adhere to any management or conservation rules and thereby steal their catches. It has been estimated that pirate fishing accounts for up to 20% of the global catch. Being so prevalent, pirate fishing is both a threat to marine biodiversity and to achieving sustainable fisheries. Pirate fishing is widespread on the high seas and, in coastal waters, it is focused on those developing countries who cannot afford to police their waters by implementing monitoring, control and surveillance measures.

Tuna Ranching in the Mediterranean Sea: In 1999 it was estimated that northern bluefin tuna (*Thunnus thynnus*) in the Mediterranean had decreased by 80% over the previous 20 years. This was due to overfishing, including pirate fishing. Since then, pressure on stocks has only increased further, not least to supply the growing number of tuna ranches in the Mediterranean. It has been suggested that without immediate action on the current level of fishing, commercial extinction of northern bluefin tuna in the Mediterranean could be fast approaching.

Freedom for the Seas: The international waters of the high seas represent the greatest remaining global commons. However, at present, the high seas are being unsustainably plundered. One of the greatest threats to deep-sea biodiversity is bottom trawling. It is suggested that new, international regulation is needed to manage the high seas using an ecosystem-based approach. In so doing, current presumptions which favour freedom of the seas and freedom to fish would be replaced with a new concept – freedom for the seas.

M. Allsopp et al., *State of the World's Oceans,*
© Springer Science+Business Media B.V. 2009

Unfair Fisheries: There is a growing concern that countries of the North are negotiating unfair and unsustainable fisheries agreements with more southerly developing coastal States who are in desperate need of quick cash. However, peoples from these coastal States are often dependent on their local fisheries as an important source of food.

Trade Liberalisation: At the World Trade Organization, five fish-exporting countries have been arguing for the elimination of all tariffs (on fish and fish products) because such trade liberalisation would benefit them economically. However, developing countries in Africa, the Pacific and the Caribbean are concerned they would lose their current trade advantages if trade liberalisation went ahead. The matter is of great concern not least because there are already real-life examples which show that trade liberalisation has led to the plundering of fish stocks, loss of marine biodiversity and undermining of food security.

Keywords Ocean governance, sustainability, IUU fishing, tuna ranching, trade, inequity, policy.

6.1   Illegal, Unreported and Unregulated Fishing (Pirate Fishing)

Illegal, unreported and unregulated fishing, also known as IUU fishing or 'pirate fishing', takes place in every ocean across the world and represents a serious global problem. Operating outside of any management and conservation rules, pirate fishers effectively steal fish both from coastal waters and the high seas, including the Southern Ocean (see Fig. 6.1). Some developing countries, in particular, lose out from the impacts of pirate fishing within their exclusive economic zones (EEZs) (High Seas Task Force 2006). The problems of pirate fishing and the nature of some of Greenpeace's campaigns on this issue in different regions are outlined below.

6.1.1   Introduction

It has been estimated that pirate fishing accounts for up to 20% of the global catch or, in monetary terms, is worth US$4 to US$9 billion per year (Greenpeace 2006e; High Seas Task Force 2006). Of this, about $1.25 billion is estimated to originate from exploitation of the high seas, and the remainder comes from the EEZs of coastal states (High Seas Task Force 2006). Because pirate fishing is so prevalent and widespread, and does not respect fishing practice regulations, it is a threat to marine biodiversity and an obstacle to achieving sustainable fisheries (Environmental Justice Foundation 2005a; High Seas Task Force 2006).

Fig. 6.1 Illegal longliner Salvora, its name obscured and crew masked by balaclavas, attempting to fish Illegally in Southern Ocean (Greenpeace/Gavin Newman)

The underlying causes of pirate fishing worldwide can be attributed to over-capacity in the world's fishing fleets and consequent competition in the fishing industry. As industrialised countries see their fish stocks decrease and impose stricter control measures in their waters, some fishing companies and boat owners find ways to evade the constraints and carry on fishing. They move their activities to areas where effective control is absent (Greenpeace 2001) or find new ways of evading the regulations. Consequently, developing coastal states who do not have the financial or technical means of effectively policing their waters by implementing monitoring, control and surveillance (MCS) measures and who may not have comprehensive, enforceable domestic fisheries laws, are often particularly hard hit by pirate fishing.

The pirates ignore or break the rules by operating without a licence and often fly flags of convenience to hide their true origins. These flags can be bought easily over the internet from several countries who ask no questions about the legality of their fishing practices. Pirates also launder stolen fish by transhipping at sea – a process by which cargo is transferred onto other vessels.

Pirate fishing, like any rampant over-fishing practice, has contributed to extensive ecological damage to marine ecosystems as well as to the target stocks (Bours and Losada 2007). Pirate fishing jeopardises the livelihoods of local fishing communities, threatens food security of developing coastal states, where there is often a dietary dependence on seafood as a source of protein, and results in significant economic losses. In addition, trespass by pirate trawlers in coastal areas has resulted in collisions with local canoes whereupon fishermen have been killed (Greenpeace 2001).

Greenpeace is campaigning to expose pirate fishing in West Africa, the Pacific and in other areas of the oceans and to implement global, legally binding measures to stop this practice completely.

6.1.2   Pirate Fishing Rampant in West Africa

West Africa depends on fish for food and income but it is the only region in the world where fish consumption is falling. Tragically, this results from the dramatic over-capacity of industrial distant water fishing fleets exploiting the region and the widespread practice of pirate fishing by these fleets.

There have been reports of illegal fishing activities off the coasts of African nations for many years. Here, pirate fishing takes advantage of some countries' inability to patrol their waters and the lack of effective controls in European ports and markets to which the fish are often destined.

In 2001, Greenpeace investigated pirate fishing in the waters off Guinea Conakry, an area where there is an absence of regulatory control. A survey of fishing activity in the region by aircraft identified 32 fishing vessels and two reefers (transport vessels). About half of the vessels were large industrial trawlers which had no visible name, flag, or other form of identification. Some vessels were even operating within the inshore exclusion zone, preserved for small canoes. Clearly, pirate fishing was rife in the region.

A further Greenpeace expedition to Guinea Conakry was undertaken in 2006 to investigate whether anything had changed as a result of the problem being previously brought to international attention and a series of measures implemented by the international community in recent times. Together with the Environmental Justice Foundation, Greenpeace revealed that, of 92 fishing vessels operating in the area at the time, about half were engaged in or linked to illegal fishing activities – including fishing without a licence, operating with no name or hiding their identity, trawling inside the 12 nautical mile zone restricted to local fishermen or transhipping at sea (the practice of transferring cargo outside of port, which is illegal under Guinean law) (Bours and Losada 2007). One such vessel found fishing illegally is shown in Fig. 6.2. Thus, little had changed since 2001. As one of the world's poorest nations, which has coastal communities dependent on fisheries, the plunder of coastal resources by pirate fishing vessels comes at an enormous social and economic cost. It has been estimated that Guinea is losing in excess of 34,000 t of fish every year to illegal fishing, worth an estimated US$110 million (Environmental Justice Foundation 2005b). Furthermore, there are indications of alarming decreases in the abundance of commercial target species, some of which are fundamental to the survival of coastal communities in the region. Pirate fishing is a threat in this regard because it increases pressure on fish stocks. The Greenpeace expedition also revealed that the pirate vessels

Fig. 6.2 The cargo vessel Binar 4, full of fish taken illegally from Guinean waters, intercepted and identified by Greenpeace and the Environmental Justice Foundation in 2006 (Greenpeace/ Pierre Gleizes)

themselves were often in a terrible condition, with crew-members subjected to near-slave conditions, exploited by the companies operating the vessels (Bours and Losada 2007).

It is known that one of the main ways vessels are able to land illegally caught fish world-wide is by transhipping their catch at sea to reefers rather than directly offloading them in ports. Following the transhipment witnessed by Greenpeace in Guinean waters in 2006 (Bours and Losada 2007), the reefer then headed for the port of Las Palmas de Gran Canaria (Canary Islands, Spain). The port of Las Palmas is well known for supporting flag of convenience fishing fleets that operate in the Atlantic Ocean (Greenpeace 2001). A report by the Environmental Justice Foundation in 2005 revealed that a number of fishing boats observed in Las Palmas had a history of illegal fishing in the waters of Guinea and Sierra Leone. This suggested that illegally caught fish is marketed to the European Community via Las Palmas port (Environmental Justice Foundation 2005b). After observing the transhipment of fish to the reefer at sea and following it to Las Palmas, Greenpeace presented a dossier of information to the Guinean and Spanish governments which proved the fish on the reefer was stolen and demanded the cargo be confiscated and returned to Guinea (Greenpeace 2006f). In response, Spanish officials agreed to declare that the fish was illegal and Guinean officials fined the owners US$150,000 (Bours and Losada 2007; Greenpeace 2006f). This incident served to illustrate the far wider problem of fish stolen from West African waters, destined for the European market.

6.1.3   Pirates Plundering the Pacific

Pacific tuna is threatened by overfishing, due to activities by foreign fishing nations including Japan, Taiwan, China, Korea, the US and the EU. Two key Pacific tuna stocks, bigeye tuna (*Thunnus obesus*) and yellowfin tuna (*Thunnus albacares*), could face commercial extinction within 3 years as a result of overfishing if no significant fishing effort reductions are introduced (Greenpeace 2006g). The problem is exacerbated by pirate fishing because there is currently a lack of resources to effectively manage and patrol most Pacific waters. In 2006, Greenpeace, together with fisheries authorities from the Federal States of Micronesia and Kiribati, undertook an expedition to inspect fishing vessels in the EEZs of both nations. This study revealed that IUU fishing in the Pacific was not necessarily associated with unlicensed vessels, but instead there was widespread under-reporting of catches and evidence of transhipments at sea. As a way of stopping IUU fishing in the region, Greenpeace has put forward a series of recommendations to the Western and Central Pacific Fisheries Commission, the organisation that was set up in 2004 to manage and protect the tuna stocks of the region. Recommendations included a ban on all at-sea transhipments, measures to prevent under-reporting of catches and measures to help conserve tuna stocks. These tuna conservation measures would include a 50% fishing effort reduction and the establishment of a network of marine reserves, including the closure of so-called 'donut holes' (pockets of high seas areas between the EEZ's of island countries) to fishing activities in order to curb the unregulated and illegal activities taking place in these areas (Greenpeace 2006g).

6.1.4   Daylight Robbery in the High Seas

Pirate fishing is particularly rampant on the high seas where large areas of oceans are still completely unmanaged. Even in areas ostensibly covered by Regional Fisheries Management Organisations (RFMOs), very little capacity monitoring, control and surveillance are carried out (Greenpeace 2005c). Of particular concern is bottom trawling, a practice which is highly destructive to marine ecosystems (see Chapter 2, Section 2.5.1). In a study of Europe's bottom-trawling fleet, Greenpeace calculated that at least 398 European vessels were equipped to engage in high seas bottom-trawling (Greenpeace 2006h). Figure 6.3 shows the proportion of vessels flagged to European countries.

Greenpeace reported that bottom-trawl fishing on the high seas is continuing largely unregulated and, at times, illegally (Greenpeace 2006h). Greenpeace has put together case studies which identify and expose IUU vessels involved in bottom trawling (Greenpeace 2005a, b). Because bottom trawling is so destructive to marine ecosystems, Greenpeace is calling for a UN General Assembly Resolution to place a moratorium on high seas bottom-trawling (see Chapter 2, Section 2.5.1.4).

In addition, remedies to counter IUU fishing are needed (see Section 6.1.5) together with the establishment of regulatory and control measures for the high seas

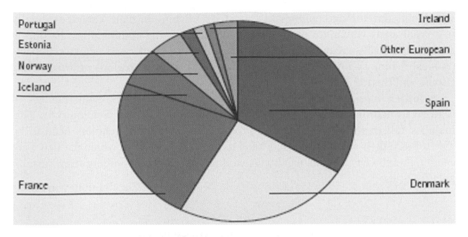

Fig. 6.3 Proportion of high seas bottom trawlers flagged in European Countries (Greenpeace 2006h)

(see Section 6.3). This would include the establishment of a central monitoring, control and compliance authority for all vessels active on the high seas. This organisation would need to operate a centralised vessel monitoring system (VMS) for all vessels licensed to fish on the high seas to enable States to distinguish between vessels fishing on the high seas from those fishing in an EEZ (Greenpeace 2006i).

6.1.5 Solutions

Trade in stolen fish by pirate vessels is fundamentally unfair and unsustainable, in many cases robbing the world's poorest people of food and income as well as plundering the marine environment. Greenpeace is campaigning to make piracy history, so as to ensure a real future for coastal communities in Africa and throughout the world. To achieve this, Greenpeace is asking governments to close ports to pirate fishing, close markets to pirate-caught fish, prosecute companies supporting pirate fishing and outlaw flags of convenience. It is purely a matter of political will to deliver the kind of enforcement that is needed to do this and so protect the marine environment and the communities that depend on it. There are already some international agreements in place that would provide comprehensive and effective measures against IUU fishing if they were properly implemented by all States. They are the United Nations Food and Agriculture Organisation (FAO) compliance agreement, the UN fish stocks agreement, the FAO model scheme for port control and the FAO international plan to prevent, deter, and eliminate IUU fishing.

With solutions already apparent, what is now urgently needed are efforts by the international community to put the necessary resources and schemes in place which will enable developing countries to curtail IUU fishing in their waters, especially in coastal areas which are of vital importance to fishing communities. Failing to act will inevitably mean the extinction of some fish stocks and negative impacts on local fishing communities in the most severely affected regions of West Africa (Bours and Losada 2007).

A further measure being taken by Greenpeace is to work with retailers of seafood to help them adopt a sustainable seafood policy (see Section 6.6). This includes full traceability of seafood products, meaning that retailers need to be able to trace their seafood products back to the very boat from which the fish were caught, thereby ensuring its legality as well as encouraging sourcing from sustainable fisheries.

6.2   Tuna Ranching and Pirate Fishing: Wiping Out Tuna in the Mediterranean Sea

The present level of fishing for northern bluefin tuna (*Thunnus thynnus*) in the Mediterranean threatens the future of this species in the region and the futures of hundreds of fishermen. There is a danger that commercial extinction of the species is just around the corner.

In May 1999, Greenpeace released a report denouncing the depletion of bluefin tuna in the Mediterranean (Gual 1999). It noted that the spawning stock biomass (total weight) of tuna was already estimated to have decreased by 80% over the previous 20 years. In addition, huge amounts of juvenile tuna were caught every season. Greenpeace reported that the main threat to the bluefin tuna at that time was IUU fishing. Seven years on in 2006, further analysis of the situation was undertaken by Greenpeace (Greenpeace 2006j), showing that threats to the tuna had worsened. Pirate fishing is continuing, but there is now an additional incentive of supplying tuna to an increasing number of tuna ranches in Mediterranean countries. While, in the past, countries from outside the region were mainly responsible for pirate fishing, these days it is vessels from within the region which are the main culprits. Tuna are caught live and taken to these ranches, where they are fed and fattened before being killed and exported, mainly to Japan. Today, due to poor management, nobody knows the exact amount of tuna taken from the Mediterranean Sea each year, but it is clear that current catch levels are well above the legal quota (Greenpeace 2006j). Figure 6.4 shows Greenpeace divers communicating these concerns using an underwater banner.

6.2.1   Quotas

The International Commission for the Conservation of Atlantic Tunas (ICCAT) is responsible for the conservation of tunas and tuna-like species in the Atlantic Ocean and adjacent seas, including the Mediterranean. In 2002, ICCAT ignored warnings

Fig. 6.4 Greenpeace divers off Menorca, Spain, hold a banner to highlight the plight of the blue-fin tuna in the Mediterranean, 2006 (Greenpeace/Gavin Newman)

by scientists that 'current catch levels were not sustainable in the long term' and adopted an unsustainable quota of 32,000t for the years 2003–2006 for the eastern bluefin tuna stock (tuna taken largely from the Mediterranean). Based on figures for catches in 2005, Greenpeace estimated that over 44,000t may have been caught in the Mediterranean, which is 37.5% over the legally sanctioned catch limit and, disturbingly, 69% above the scientifically recommended maximum catch level (Greenpeace 2006j). More recently, the catch was estimated to be over 50,000t (Losada 2007). The 2006 Atlantic Bluefin Tuna Assessment Session of the Scientific Committee on Research and Statistics of ICCAT, which took place in Madrid in June 2006, stated that *"The volume of catch taken in recent years likely significantly exceeds the current Total Allowable Catch and is likely close to the levels reported in the mid-1990s, i.e. about 50,000t in the East Atlantic and the Mediterranean"* (SCRS 2006). This high level of piracy in the region is a clear threat to tuna stocks and cannot be sustainable.

6.2.2 Ranching and Pirate Fishing

Currently, most of the bluefin tuna catch in the Mediterranean goes to tuna ranches, following capture and transfer to large cages for transport (see Fig. 6.5 and Plate 6.1). Tuna ranching began in the late 1990s; there has been little or no regulation of the industry and ranching has boomed, spreading to 12 countries

Fig. 6.5 Turkish tuna fleet purse seine fishing and transferring catch to transport cage (Greenpeace/ Gavin Newman)

Plate 6.1 Bluefin tuna swim inside a transport cage, Mediterranean (Greenpeace/Gavin Newman)

1985	1996	2000	2001	2002	2003	2004	2006
Spain	Spain	Spain	Spain	Spain	Spain	Spain	Spain
	Croatia	Croatia	Croatia	Croatia	Croatia	Croatia	Croatia
		Malta	Malta	Malta	Malta	Malta	Malta
			Italy	Italy	Italy	Italy	Italy
				Turkey	Turkey	Turkey	Turkey
					Cyprus	Cyprus	Cyprus
					Libya	Libya	Libya
						Greece	Greece
						Lebanon	Tunisia
							Morocco
							Portugal
							Lebanon

Fig. 6.6 Tuna Farming proliferation in the Mediterranean and East Atlantic (Lovatelli (2005) and ICCAT database on declared farming facilities [www.iccat.es/ffb.asp])

by 2006 (see Fig. 6.6). The total reported farming capacity of the ranches, at 51,012 t, far exceeds the total allowable catches set by ICCAT of 32,000 t. This is an indisputable incentive for illegal fishing in the region. Indeed, an examination of available trends in the industry clearly suggests that illegal fishing is supplying ranches (Greenpeace 2006j).

6.2.3 Greenpeace Recommendations

Without immediate action, the future of the northern bluefin tuna is in jeopardy in the Mediterranean Sea. Key to restoring the population are (1) measures to protect important sites where tuna congregate to breed and feed, by establishing marine reserves, (2) improvement of the fishery management for the whole Mediterranean, and (3) prevention of the further expansion of tuna ranching:

- *Marine Reserves*
Marine Reserves are areas of the sea that are fully protected from damaging human activities including fishing. Although migratory species like the bluefin tuna do not spend all their time in any one area, they can be protected by marine reserves at critical sites such as breeding grounds. Large-scale marine reserves are necessary to protect bluefin tuna in their breeding grounds, like the Balearic Islands.

- *Precautionary Management*
Mismanagement is widespread in the bluefin tuna fishery in the Mediterranean. For instance, misreporting and under-reporting are common in the area. Greenpeace suggests that measures which should be adopted by ICCAT to manage the fishery sustainably include: (1) a substantial reduction of the bluefin tuna quota as part of a long-term tuna rebuilding programme set on a precautionary basis; (2) a new minimum landing size that matches the sexual maturity of the species; and (3) independent observers both on board tuna fishing vessels and in the ranches to record and report the catch, to ensure that under-sized fish are not caught, the quota is not exceeded, and that information is available to manage the fishery sustainably.

- *Stop Further Ranching*

The expansion of tuna ranching must be stopped until the northern bluefin tuna population recovers and the fishery is properly managed.

6.2.4 An Unsustainable Future?

At a meeting of ICCAT in November 2006, a handful of delegations called for better regulation of the number of tuna caught per year but this was prevented by those Mediterranean countries, and other countries, more involved in the fishery. ICCAT even failed to agree to stop fishing during the spawning season and blocked proposals for the creation of a working group to identify those responsible for illegal catches. Such poor management of the fishery offers no room for the recovery of the northern bluefin tuna population. It is a clear example of the failure of governments and Regional Fisheries Management Organisations (i.e. ICCAT) to guarantee the sustainable management of the marine resources for which they claim to take responsibility.

6.3   Freedom for the Seas

Nowadays, due to advances in technological and industrial efficiency, fishing vessels are able to fish across the globe and to depths of two kilometres (see Chapter 2). On the high seas (in regions beyond Exclusive Economic Zones), anyone who has the money, the technology and the inclination can basically fish for whatever they want (Greenpeace 2005d). This is because fishing and scientific exploration on most of the high seas is not subject to any regulation. Consequently, a situation has arisen whereby marine biodiversity is being unsustainably plundered because of the legal gaps and the lack of political will to change the *status quo* (Greenpeace 2005c). Unless urgent action is taken, future generations will be denied the chance to experience or enjoy the benefits of the life that thrives within the international waters of the high seas - the greatest remaining global commons (Greenpeace 2005d). In 2005, Greenpeace published three reports which discussed the lack of regulation on the high seas and, importantly, proposed solutions to bring about sustainable and equitable regulation (Greenpeace 2005c, d; McDiarmid et al. 2005). A summary of the main points of these reports is discussed here.

The largest and most immediate threat to deep-sea biodiversity on the high seas is the practice of bottom-trawling (see also Chapter 1, Section 1.2.1 and Chapter 2, Section 2.5.1). Industrial fishing vessels such as deep-sea bottom trawlers are laying waste to parts of the deep which are oases of marine life. In just one sweep, a single bottom trawl can leave areas of the surface of a seamount almost devoid of life (Greenpeace 2005c).

Also unregulated on the high seas is the expanding exploration of deep-sea marine life for scientific as well as commercial purposes, a practice known as bioprospecting (McDiarmid et al. 2005; see also Chapter 1, Section 1.2.3). Bioprospecting generally involves the sampling of marine plants, animals or microorganisms for scientific research and may be followed by their commercial harvesting if the unique biochemicals they contain are useful for health or other commercial purposes. To date, bioprospecting has mainly taken place in more shallow marine waters, although scientists are beginning to appreciate the valuable resources that are found in the depths of the high seas. The dangers posed by bioprospecting include physical disturbance or disruption of ecosystems, potential pollution and contamination, and problems of over-harvesting.

Greenpeace believes that high seas marine biodiversity forms part of the global commons, a heritage for all humankind, which must be managed in a way to ensure its long-term sustainability. What is needed to fill the present legal void in regulation is an integrated, precautionary and ecosystem-based management approach to promote the conservation and sustainable management of the marine environment in areas beyond national jurisdiction, including equitable access and benefit sharing of these resources. By adopting these measures, current presumptions which favour freedom of the high seas and freedom to fish would be replaced with the new concept of freedom for the seas (Greenpeace 2005c, d; McDiarmid et al. 2005).

Greenpeace also believes that one effective way to bring about the more sustainable and equitable management for the high seas would be via the adoption of a new Agreement under the auspices of the United Nations Convention of the Law of the Sea (UNCLOS). The Law of the Sea Convention not only offers States the right to use the oceans but also requires States to fulfil numerous duties including taking measures as may be necessary for the protection and preservation of the marine environment (Greenpeace 2005b). What is now urgently needed to protect high seas biodiversity is a new UNCLOS agreement to implement those obligations and to fill the regulation void. This new implementing agreement should include the following requirements:

- Provide a clear mandate and legal duty to protect biodiversity on the high seas, founded on ecosystem-based management and the precautionary principle.
- Establish an effective, centralised monitoring, control and surveillance (MCS) mechanism for human activities on the high seas. This would need to have legal 'teeth' to ensure that activities undertaken comply with international law. It could be funded, for instance, by dues paid by States for the number of vessels authorised to undertake extractive activities on the high seas.
- Give a clear mandate for the identification, selection, establishment and management of high seas marine reserves. Marine reserves are areas closed to all extractive uses such as fishing and mining, as well as to industrial and disposal activities. Greenpeace is calling for a global network of marine reserves, covering 40% or more of the oceans, to help protect marine biodiversity (see also Chapter 7).
- With regard to bioprospecting, it is necessary that an environmental impact assessment be carried out before approval is given for activities planned for the ocean floor or in the high seas.
- Encourage information and knowledge-sharing on high seas biodiversity through the creation of a central list of high seas species available to all.
- Establish a regime for benefit-sharing.

All of these components for a new UNCLOS implementing agreement are essential to ensure that the living resources of the high seas' global commons are sustainably and equitably managed, now and for the future (Greenpeace 2005c, d; McDiarmid et al. 2005). While such an Agreement is being negotiated, States must reach agreement on the key criteria for the establishment of high seas marine reserves covering 40% of our oceans. They must also begin to identify such areas (Roberts et al. 2006) and put an immediate halt to destructive fishing practices such as high seas bottom trawling.

6.4 Unfair Fisheries

Due to the problems of over-fishing and depleted stocks in their own waters, fishing fleets of the North are now turning their attention further south to the waters of developing coastal states such as those off West Africa and the Pacific Island States. In these regions, local people are often dependent on seafood as an important food source and small-scale fisheries provide livelihoods for many. As for the case of the Pacific Island States, fisheries are one of their only abundant natural resources and, as such, sustainable and equitable fisheries are crucial for their economies. As foreign industrialised fishing fleets move in on their waters, which, due to lack of resources and infrastructure are often poorly managed and monitored, there is an increasing threat that fisheries will be overwhelmed and stocks decimated. This would ruin coastal fishing communities and leave local people without income and a source of nutrition that is often so vital for their survival.

In addition to the threat of over-fishing, the movement of industrialised fleets to the South has been financially exploitative and access agreements are sadly often in the hands of private companies that negotiate 'sweetheart' deals with the sometimes corrupt governments. For example, with regard to tuna fishing in the Pacific, the economic return from access fees and licences paid by foreign fleets is a mere 5% or less of the US$2 billion the fish is worth (OXFAM 2006). On top of this there is further pressure on fish stocks from illegal, unregulated and unreported (IUU) fishing.

To promote solutions to these problems, Greenpeace is working to achieve sustainable and equitable fisheries by bringing an end to the unfair and unsustainable fisheries agreements that the distant water fishing nations negotiate with the developing coastal states which are in desperate need of quick cash. By negotiating fairer deals, the coastal states can manage their resources in a sustainable way, ensure continued livelihoods and income for their coastal communities and, with the assistance of the distant water fishing nations, build up capacity to gain the economic and social benefits from their natural resources in a sustainable manner.

6.5 Trade Liberalisation (Free Trade) Means Empty Oceans

At present, worldwide trade in the fishing industry comes under the control of a very complex system of tariffs for fish and fish products (Allain 2007). The purpose of tariffs is to protect a country's domestic interests where they would be vulnerable to foreign competition if the tariff protection was not there. In the case of fish and fish products, those interests are distant water fleets, domestic harvesters or fish processors who could easily be undermined by foreign competition. As part of the negotiations at the World Trade Organisation (WTO), five fish-exporting countries have been arguing for the elimination of all tariffs because such trade liberalisation would benefit them economically (at least in the short term). However, developing countries in Africa, the Pacific and the Caribbean are concerned that they would lose their current trade advantages if trade liberalisation went ahead.

The WTO was created in 1995 to replace the General Agreement on Tariffs and Trade (GATT), which started the process of trade liberalisation after the Second World War. In November 2001, after years of discussions and concerted pressure exerted on developing countries, WTO members agreed to launch a series of multi-faceted trade negotiations known as the Doha Round. This committed governments to undertake negotiations in non-agricultural market access (NAMA), which included fish and fish products. However, recent NAMA negotiations have essentially been a tug-of-war over tariffs, fought between developed and developing countries. Some developed countries sought deep and rapid tariff cuts under a 'zero for zero' scenario whereby they proposed to cut their tariffs to zero and expected developing countries to do the same. However, tariffs are often the last industrial policy instruments left to developing countries. In the NAMA negotiations, developing countries by and large resisted the push for the zero for zero scenario and sought to maintain tariff protection for future industrialisation and to protect vulnerable sectors of their economies in terms of livelihoods and food security.

If the Doha Round succeeded in bringing about further trade liberalisation in the fishing industry, what would be the consequences? A theoretical study, carried out by the Committee for Fisheries of the Organisation for Economic Co-operation and Development (OECD) and published in 2003, predicted that trade liberalisation could lead to the over-exploitation of stocks and catch declines for both fish exporting and importing countries. This is of great concern because, according to FAO figures, 77% of fish stocks worldwide are already either fully exploited – i.e. with no room for further expansion – or in even worse shape than this: over-exploited, depleted or recovering from depletion (see Chapter 2, Section 2.2.1).

Aside from such predictive theory, there are real life examples which demonstrate that trade liberalisation causes many problems due to the resulting export increases or opening of waters to foreign fleets. For example, liberalisation of the fishing trades in Mauritania and Argentina, where trade agreements were made with the EU, led to serious over-fishing, plundered stocks and a reduction of marine biodiversity. In Senegal, free trade with the EU market led to over-fishing and, as a country dependent on fish for 75% of the population's animal protein needs, an undermining of the country's food security.

From both the theoretical study and practical examples, the Greenpeace report on trade liberalisation (Allain 2007) concluded that when the fish trade is liberalised in a context of deficient management, or, worse, no management at all, it quickly leads to the over-exploitation of fisheries resources, social harm and environmental degradation. For developing countries, there is a likelihood that trade liberalisation

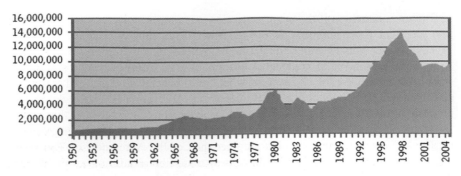

Fig. 6.7 Fish landings in Argentina (tonnes) from 1950 to 2004 (Greenpeace 2007a)

will deplete their fish stocks and result in a loss of biodiversity. For example, trade liberalisation in the 1990s in Argentina resulted in steep increases in catches (Fig. 6.7) and by 2002 at total of six Argentinean fish stocks were considered endangered. For the local consumers of developing countries where liberalisation occurs, fish prices will rise as more national fishing effort is diverted to fish for export species, leading to a lesser supply of locally fished and consumed pelagics.

If the Doha Round's NAMA negotiations were to succeed in further trade liberalisation, this would only bring economic benefits to a handful of developed, fish-exporting countries, and fish stocks would come under increasing pressure. As of July 2006, the entire Doha Round was suspended. Greenpeace believes that this is an opportunity to avert the potential disasters of trade liberalisation by taking the negotiations on fish and fish products out of the WTO and putting them where they should have been all along: in multilateral fora, where commercial and trade interests do not dominate and where, ideally, sustainability and the protection of the environment are the focus of discussions. In this regard, States should take action to ensure that existing international law is implemented fully. Legal instruments which are largely ignored, but which require countries to police themselves and fleets and to fish responsibly, include the UN Fish Stocks Agreement, the FAO Code of Conduct for Responsible Fisheries, and the World Summit of Sustainable Development's Johannesburg Plan of Implementation. At the very least, until such time as these instruments are universally adhered to and enforced, it would be irresponsible for the members of the WTO to engage in further liberalisation on fish and fish products.

6.6   Towards Sustainable Fisheries

In simple terms, a particular seafood is sustainable if it comes from a fishery whose practices can be maintained indefinitely without reducing the target species' ability to maintain its population, and without adversely impacting on other species within the ecosystem directly, by removing their food source or by damaging their physical environment (Dorey 2005). On the basis of these basic criteria, most current world fisheries cannot be considered to be sustainable (see also Chapter 2).

The fishing industry itself can be slow to implement moves towards sustainability. A complementary approach – a demand in the marketplace for sustainable seafood – can act from the bottom up to pressurise fisheries to move towards sustainability. Greenpeace has taken action at the level of the marketplace by making major seafood retailers aware of unsustainable fishing practices and by providing recommendations on the sourcing of sustainable seafood. By way of an example, this section discusses actions taken by Greenpeace UK with major seafood retailers. Other action taken in Sweden has been to raise awareness of the problem of illegally caught fish and stop it reaching the marketplace.

6.6.1   Greenpeace Action in the UK

In the UK in 2004/5, the Seafood Industry Authority reported that over 85% of chilled and frozen seafood sales were through supermarkets. Because the supermarkets control such a high fraction of seafood sales, the supermarkets hold a vital key to improving the sustainability of fisheries. Consequently, Greenpeace UK conducted research into the sustainability policies of UK supermarkets and published the results of the study in 2005 (Greenpeace 2005e).

This Greenpeace report showed that most UK supermarkets purchased seafood with little consideration for the health of fish stocks and with even less consideration for where it was caught or for the impacts on the wider marine environment. Labeling practices were generally poor in that they did not help the consumer to make choices on the sustainability of the seafood by providing the species name, the precise area of catch or the fishing method used. Overall, the results of the Greenpeace study clearly showed that UK supermarkets were not rising to the challenge of sourcing sustainable seafood. To remedy the situation, Greenpeace developed a model sustainable seafood policy which it asked supermarkets to adopt. Briefly, the recommendations were to:

- Remove the worst – immediately stop the sale of overfished stocks and those caught using wasteful or destructive methods.
- Support the best – increase the range of sustainable seafood sold from stocks that are not depleted and using the most sustainable fishing methods.
- Change the rest – by working with suppliers to source fish from the least depleted stocks, and working with the fishing industry and/or researchers to improve the sustainability of fishing methods. Rejecting fish from fisheries and suppliers that refuse to change.
- Improve seafood labeling – so that consumers can make a more informed choice about the sustainability of the products they are buying.

One year after the report was released, Greenpeace investigated whether UK supermarkets had risen to the challenge of sourcing sustainable seafood, and published their findings in a report (Greenpeace 2006k). Fortunately, most of the supermarkets had improved their policy on sourcing sustainable seafood. While this is very encouraging, there is still more work to do. For example, whilst improvements in own brands had been made, supermarkets continued to sell other branded seafood products that were not sustainable because they were outside their sustainable sourcing policies. This has to change. In addition, Greenpeace UK has made demands on supermarkets to take action to radically improve the UK's skate fisheries, or to stop selling skate, because these fish are often caught by destructive fishing methods, the fisheries are poorly managed and the stocks are in decline. Greenpeace also made demands on supermarkets to stop selling species that are caught by a highly destructive bottom-trawling practice known as beam trawling (see further Chapter 2, Section 2.5.2).

As a guide to consumers, Greenpeace UK has produced a league table of the major supermarkets to highlight their performance with regard to sustainable seafood sourcing. They are also inviting consumers to email those supermarkets that are under-performing to bring about pressure for positive change towards sustainability.

6.6.2   Greenpeace Action in the Baltic

In 2006, Greenpeace released a report which highlighted the unsustainability of the cod fishery in the Baltic and the problem that roughly a third of the cod are taken by illegal, unregulated and unreported fishing (pirate fishing) (Greenpeace 2006l).

In the Baltic Sea, the western cod stock (subspecies *Gadus morhua morhua* L.) has been much reduced over the past 3 decades and the eastern stock (subspecies *Gadus morhua callaris* L.) has been hit extremely hard by overfishing. Despite the poor shape of the eastern Baltic stock, the Council of Fisheries Ministers of the EU set a catch level in 2006 of more than three times that advised by the International Council for the Exploration of the Sea (ICES: the scientific advisory body for the north-eastern Atlantic region). In addition to this legal catch, a huge amount of illegally caught cod is landed in harbours around the Baltic Sea for consumption within the EU market. For instance, in 2005, ICES estimated that 38% of cod in the eastern Baltic Sea was taken illegally.

Following publication of the Greenpeace report on Baltic cod, some of the large supermarkets in Sweden declared that they would stop selling Baltic cod, a decision that was subsequently matched by the remaining supermarkets (Greenpeace 2007a). Greenpeace also provided the means for the general public to e-mail some of the biggest buyers of Baltic cod to persuade them to stop buying it.

Chapter 7
Marine Reserves

Abstract Marine wildlife is under threat from many human activities, especially overfishing. Conservation measures, together with sustainable and equitable management, are now urgently needed to help restore and protect ocean biodiversity. Marine reserves are increasingly being proposed by scientists and policy-makers as a tool for the conservation and management of the oceans. Fully protected marine reserves protect entire areas from harmful human activities and, in so doing, protect the full variety of species and their habitats within ecosystems, a prerequisite for conserving biodiversity. Marine reserves can therefore be seen as national parks of the sea.

Marine reserves have been shown to have many benefits. They result in long-lasting and often rapid increases in the abundance, diversity and productivity of marine organisms. They can also benefit fisheries adjacent to reserves as a result of the spill-over of fish across the reserve boundaries and spill-over of larvae or eggs. Marine reserves also provide areas for research and discovery and well managed, non-destructive tourist activities such as scuba diving and snorkeling. Although marine reserves cannot directly deal with the impacts of transboundary pollution or climate change, their ecosystems become more resilient than those of exploited areas and this could help to mitigate some of the negative consequences.

Some marine reserves have been created recently but the speed at which new areas are being designated and protected is presently too slow. Only about 0.1% of the oceans are covered by fully protected marine reserves. Greenpeace is calling for 40% of the oceans to receive such protection. Greenpeace proposes establishing a representative network of marine reserves comprising large-scale reserves on the high seas and a mosaic of smaller marine reserves within the coastal zone.

Keywords Conservation, protection, restoration, marine reserves, MPA, planning, site selection, data generation, reserve networks, connectivity.

7.1   Introduction

Previous chapters in this book have described some of the major threats facing marine wildlife. Pressure from overfishing, pirate fishing and the use of destructive fishing techniques has caused harm to many species and habitats. The United Nations Food and Agricultural Organisation has estimated that 77% of fish stocks worldwide are already either fully exploited – i.e. with no room for further expansion – or in even worse shape than this: over-exploited, depleted or recovering. Fishing on an industrial scale has caused catastrophic declines in some of the world's predatory fish, such that stocks have been reduced by 90% of their pre-industrialised fishing levels (e.g. Christensen et al. 2003; Myers and Worm 2003; see Chapter 2, Section 2.3.2). Some fishing techniques, for example, bottom trawling, are highly destructive to marine habitats and ecosystems. Bottom trawling has stripped seamounts (underwater mountains) of rich and delicate coral and sponge forests that may have taken thousands of years to develop (see Chapter 1, Section 1.2.1). This reduces habitat for a myriad of other species and food for others. The recovery of impacted ecosystems can be extremely slow; it may be centuries before the scars heal (Roberts et al. 2006). Some fishing methods result in substantial numbers of marine mammals, turtles and seabirds being caught incidentally and many die as a result. This incidental take of non-target animals, known as by-catch, has been implicated in the population declines of many seabirds, marine mammals and turtles (see Chapter 2, Section 2.6).

Pressure from fishing has impacted regions both within Exclusive Economic Zones (EEZs) and beyond, on the high seas. The lack of regulation of fishing on the high seas has led to a situation where marine biodiversity is being unsustainably plundered (see Chapter 6, Section 6.3). Unless urgent action is taken, future generations will be denied the chance to experience or enjoy the benefits of the life that thrives within the international waters of the high seas – the greatest remaining global commons. Both regions within EEZs and the high seas are also subjected to threats in addition to fishing pressure, including pollution (see Chapter 4) and climate change (see Chapter 5). In coastal zones, there is an ongoing decline in many of the world's coral reefs, principally due to overfishing and pollution (see Chapter 1, Section 1.4), as well as coral bleaching due to climate change (see Chapter 5, Section 5.2.2). With regard to mangroves, it has been estimated that there has been a loss of 35% of their original area in the past 2 decades (see Chapter 1, Section 1.5). A number of human activities have caused the destruction of mangroves, a major threat being their conversion for commercial aquaculture purposes, particularly shrimp aquaculture (see Chapter 3, Section 3.2.1.2).

It is therefore clear that marine wildlife is seriously under threat from many human activities. Indeed, a recent study produced a global map of total human impact on the oceans and concluded that more than a third (41%) of the oceans are heavily affected by human activities and that no area remains unaffected (Halpern et al. 2008). Conservation measures, together with sustainable and equitable management, are now desperately needed to help to restore and protect ocean biodiversity. To conserve biodiversity means not only protecting any single species, but the full variety of species and their habitats, as well as preserving the complex interactions between species that make up an ecosystem. To do so requires an approach that considers all these aspects. Marine reserves, which protect entire areas from a range of human impacts, do just this, which makes them a unique tool for conservation (Greenpeace 2006m). Marine reserves also provide a vital tool for implementing an ecosystem-based management approach to fisheries, that is, an approach that promotes conservation and the sustainable use of living resources in an equitable way and is underpinned by the precautionary principle.

Experience of marine reserves in different areas of the world shows that they convey many benefits to ocean life, including increasing the number, size and diversity of species within the reserves. Marine reserves are also being increasingly recognised by scientists and policy makers as a key tool for the conservation and management of the oceans. Greenpeace is proposing the establishment of a global network of marine reserves covering 40% of the ocean surface. To date, Greenpeace has published three reports that propose the establishment of a network of marine reserves in different areas – a global network on the high seas (Roberts et al. 2006), a network in the Mediterranean (Greenpeace 2006m) and a network in the North and Baltic Seas (Greenpeace 2004b). Drawing on these reports, the following chapter discusses the benefits of marine reserves, the selection of appropriate sites and ways of implementing them which can ensure their effectiveness in practice.

7.2   Marine Reserves Defined

Marine Protected Area (MPA) is a term that is becoming increasingly common in the context of biodiversity conservation, habitat protection and fisheries management. The term covers a wide range of protection measures, with an equally wide variation in the benefits conferred by this status. MPAs can be created for many purposes, ranging from the protection of a single species to that of a whole habitat or ecosystem, and can encompass the protection of certain interests, such as small-scale or recreational fishing (Greenpeace 2006m).

Marine reserves are one type of MPA, and in terms of protecting the marine environment, they offer the highest level of protection. Marine reserves are areas of the sea that are fully protected from damaging human activities – much like national parks in the sea. In 2004, Greenpeace adopted the following definition of marine reserves (Greenpeace 2004b):

Large-scale marine reserves are areas that are closed to all extractive uses, such as fishing and mining, as well as to disposal activities. Within these areas there may be core zones where no human activities are allowed, for instance areas that act as scientific reference areas or areas where there are particularly sensitive habitats or species. Some areas within the coastal zone may be opened to small-scale, non-destructive fisheries, provided that they are sustainable, within ecological limits, and have been decided upon with the full participation of affected local communities.

In essence then, fully protected marine reserves are places that are protected from all fishing and other extractive or harmful human uses, such as mining and drilling for oil. They are also protected from harm by other causes, so far as it is possible, such as pollution. Recreational boating, passage of shipping etc. are permitted up to levels that do not harm the environment. Marine reserve status does not interfere with the right of innocent passage embodied in the UN Law of the Sea. However, reserves may require additional restrictions on shipping where such areas are also designated as Particularly Sensitive Sea Areas. In the case of high seas marine reserves, they are unlikely to attract much recreational use since they are generally inaccessible and lack the easily approachable animal life that shallow coastal seas often possess. The main source of harm against which high seas marine reserves must offer protection is fishing, in both its legal and illegal (IUU) forms. In future, as more accessible oil, gas and mineral resources are depleted, reserves will also need to offer protection from drilling and mining (Roberts et al. 2006).

With regard to the establishment of coastal marine reserves, the full involvement of local communities from the initial setting up process onwards will be crucial in determining their success over time. If people feel a sense of ownership for the marine resources in their area and have their views taken into account when reserves are sited, they are far more likely to support them. In this regard local fishermen are key community members who will need to be involved. They have to believe that the marine reserves are acting in their best interests and are not there to take away their livelihoods. By negotiating with fishermen, any misunderstandings about the reserves can be remedied and all their views can be properly represented (Roberts and Hawkins 2000). Pioneering work has been done in the Caribbean with the establishment of the Soufrière Marine Management Area in St. Lucia. This patchwork of closed areas of reef has sharply increased catches outside the closed areas. The areas that are open to fishing support as many fishermen as the whole reef did before the marine reserves were established. Much of the success of this network can be attributed to the full involvement of different stakeholder groups from the planning stages onwards (Renard 2001).

Comparing the benefits yielded by two different marine reserves in the Philippines, Apo Island marine reserve and Sumilon Island, underscores the importance of community involvement. The local community has given its continuous support to Apo Island and the benefits have accrued over time, whereas local politics at Sumilon resulted in a breakdown of reserve protection and a subsequent loss of reserve benefits. Lessons learned from these experiences are being broadly applied to other marine conservation projects in the Philippines, in particular the importance of community involvement (Sobel and Dahlgren 2004).

7.3   Benefits of Marine Reserves

The establishment of marine reserves has been shown to result in long-lasting and often rapid increases in the abundance, diversity and productivity of marine organisms (American Association for the Advancement of Science 2001). While the benefits of protection are more apparent for species that spend much or all of their time within a marine reserve, reserves can also offer protection to migratory species if they are protected at vulnerable life-stages, such as through the protection of spawning and nursery grounds (Greenpeace 2006m).

Marine reserves can directly address the problems of ecosystem damage, for instance, where a species has been depleted by overfishing or where habitats have been damaged through destructive activities such as bottom trawling. Marine reserves can help to restore ecosystem balance that has been lost through human activities. For example, fishing target species can upset the balance of predator/prey relationships, and result in habitat change. As these changes can occur over a long period of fishing at unsustainable levels, the altered habitat is sometimes not recognised as unnatural until a marine reserve is designated and larger fish return and restore balance. For example, a reserve was created in New Zealand waters in an area that was so heavily grazed by sea urchins that there was little seaweed left. This was due to their predators, snapper and rock lobster, being heavily fished. After 20 years of protection, these predator species increased and, in turn, the predators reduced sea urchin densities, which allowed the seaweed beds to recover (Marine Conservation Unit 2005).

While marine reserves cannot combat directly the impacts of pollution or climate change, by nature their ecosystems become more resilient than those of exploited areas and they may therefore mitigate some of the negative consequences. Changes to oceans and marine life as a result of increased emissions of greenhouse gases are likely to be highly complex, including changes in sea temperature, sea level, currents and ocean chemistry (see Chapter 5). On top of these pressures to marine ecosystems, there are also serious pressures from overfishing to the effect that many species are now severely depleted (see Chapter 2). Such negative pressures on fish stocks will be likely to exacerbate the impacts of climate change on marine ecosystems (Greenpeace 2007b). For example, cod (*Gadus morhua*) stocks in the North Sea have been considered to be at the risk of collapse since 1990 because of overfishing. In addition to this, climate change has influenced the occurrence of different plankton species, which has made it difficult for larval cod to find food. Furthermore, it seems that the reproductive success of cod is also hampered by the warmer environment. Thus climate change can add further negative influences on already over-stressed marine ecosystems. In this regard, the Secretariat of the Convention on Biological Diversity (2003) has advised that:

> *Conservation of biodiversity and maintenance of ecosystem structure and function are important climate change adaptation strategies because genetically diverse populations and species-rich ecosystems have a greater potential to adapt to climate change.*

(Secretariat of the Convention on Biological Diversity 2003)

A global network of marine reserves could help to protect marine ecosystems by acting as an insurance policy for an unpredictable future due to climate change. Marine reserves could, in other words, thus provide the necessary safety net to strengthen the resilience of marine ecosystems and limit the impacts of climate change (Greenpeace 2007b).

7.3.1 Benefits to Fisheries

Marine reserves are, by definition, closed to fisheries, but the establishment of a network of marine reserves can benefit fisheries in a number of ways: marine reserves enable exploited populations to recover and habitats modified by fishing to regenerate. There is also a growing body of evidence to suggest that the establishment of a network of marine reserves can lead to enhanced yields in adjacent fishing grounds. This can be the result of either the spillover of adults and juveniles across reserve boundaries or from the export of larvae or eggs from reserves to fished areas (Greenpeace 2006m). Also, because reserves contain more and larger fish, protected populations can potentially produce many times more offspring than can exploited populations. For example, one of the major problems with fish stocks that have been depleted by overfishing is that there are very few large fish remaining in the population. Large females are essential, because they produce many eggs of better quality; generally when a female doubles in length, she produces eight times more eggs (Roberts and Hawkins 2000). These eggs exhibit a higher level of fertilisation and better survival rates than those from less mature fish. So, a few large mature females may contribute far more to reproduction than a large number of first-time spawning females. In marine reserves, some female fish will, over time, grow large and make very significant contributions to the eggs and larvae that may be exported out of the reserve.

Improvements in fishery catches have been found adjacent to established reserves. An example of this can be seen at the Soufrière Marine Management Area along the southwest coast of the Caribbean island of St. Lucia. There, along an 11-km stretch of coastline, a network of five marine reserves was established in 1995, covering about 35% of the coral reef fishing grounds. After a period of 5 years, research was carried out to assess whether the reserves had had any impact on the neighbouring local fisheries. While the total fishing effort remained stable, there were significant improvements in the catches. For example, the mean total catch per trip for fishers with large traps increased by 46% and for fishers with small traps increased by 90%. Catch per trap increased by 36% for big traps and 80% for small traps. Thus, after a period of 5 years, the marine reserves had led to improvement in the local fishery, despite the 35% decrease in the area of fishing grounds (Roberts et al. 2001). Another example of improvements to fisheries can be found in the series of marine reserves in Egypt's Red Sea that were established in 1995. The reserves led to an increase of over 60% in the catch per unit effort of a surrounding fishery after only 5 years of protection (Galal et al. 2002).

7.3.2   Other Benefits of Marine Reserves

Monitoring the biodiversity of marine ecosystems that are protected against extractive activities has great educational value. From schools to universities and research institutions, marine reserves provide areas for research and discovery. Other activities such as diving, snorkeling, underwater photography and whale watching all benefit from diverse and abundant marine life. Such activities can provide alternative economic opportunities for coastal communities, providing they are well managed and do not compromise the level of protection conferred by reserve status. For example, tourism is the most economically valuable industry in the Great Barrier Reef Marine Park (Commonwealth of Australia 2003). Another example comes from Apo Island marine reserve in the Philippines. Here it was estimated that the initial US$75,000 investment in the reserves later yielded an annual return of somewhere between US$31,900 and US$113,000 taking into account increased fish yields outside the marine reserve, and other reserve-generated income such as increased local dive tourism (White et al. 2000).

7.4   Planning of Marine Reserves

The overarching aim of Greenpeace's proposals for a global network of marine reserves is to establish a network that is representative of the full variety of life in the sea. Such a network endures to protect places that are biologically rich, supporting outstanding concentrations of animals and plants. It also seeks to protect places that are particularly threatened or vulnerable to present or possible future human impacts, like fishing or seabed mining (Roberts et al. 2006).

7.4.1   Coverage of Marine Reserves

A number of organisations and scientists have proposed the extent of coverage needed for networks of marine reserves to be effective. For example, the UK Royal Commission on Environmental Pollution (RCEP) in their 2004 report on fisheries suggested that:

> Selection criteria should be developed for establishing a network of marine protected areas so that, within the next five years, a large scale, ecologically coherent network of marine protected areas is implemented within the UK. This should be at least 30% of the UK's exclusive economic zone being established as no-take reserves closed to commercial fishing.

(RCEP 2004)

Similarly, the World Parks Congress in 2003 recommended that at least 20–30% of all marine habitats should be included in networks of marine reserves (World

Parks Congress Recommendation 22, 2003). There are good scientific arguments
for taking an even more precautionary approach, since higher levels of protection
can be required to maintain the integrity of marine ecosystem processes. Gell and
Roberts (2003) reviewed nearly forty studies examining how much of the sea
should be protected. The majority of studies concluded that between 20% and
50% of the sea should be protected to achieve the conservation of viable populations, support fisheries management, secure ecosystem processes and assure sufficient connectivity between marine reserves in networks (Roberts et al. 2006). In
reports on marine reserves proposals for the high seas (Roberts et al. 2006), the
North and Baltic Seas (Greenpeace 2004b) and the Mediterranean (Greenpeace
2006m), Greenpeace adopted the goal of protecting 40% of the oceans.

Urgent action is now needed to implement such a coverage of marine reserves,
given the degenerating state of the oceans and because the level of implementation
to date has generally been far too slow. For example, it has taken some 30 years to
achieve the level of about 1% ocean area protection that presently exists, and only
about 0.1% is fully protected.

Although the scale and the rate of implementation of marine reserves needs to
be vastly improved, the concept of fully protected marine reserves is gaining widespread acceptance. For example, in 2006, the world's largest marine conservation
area was created off the coast of the northern Hawaiian Islands and was designated
to be a national monument. It encompasses nearly 140,000 miles^2 of U.S. waters
including 4,500 miles^2 of relatively undisturbed coral reef that is home to more than
7,000 species (NOAA 2006b). In California, a marine reserve which has designated
no-fishing zones was created in 2007 to protect 204 miles^2 of waters along the
central coast (California Department of Fish and Game 2007). In Mexico, in 2006,
the government declared the Espiritu Santo Archipelago to be a new marine reserve
(Greenpeace 2006n). In 2004, the Great Barrier Reef Marine Park Authority introduced a new zoning plan aimed at protecting the entire range of plants and animals
on the reef. Some of the zones are designated as 'no take' zones, in which fishing
and collecting is prohibited (Great Barrier Reef Marine Park Authority 2007). At
Palau, a tiny island state 600 miles east of the Philippines, overfishing by outsiders
in the 1980s and 1990s had left reef fish depleted. In 1994 elders from one village
banned fishing on a small area of reef. After a few years, the fish again became
more plentiful and other villages started to do the same. By 2007, 460 miles^2 of
reefs and lagoons on Palau have been protected and the area has become internationally known for recreational diving (Pala 2007).

7.4.2   Size of Marine Reserves

Greenpeace proposes establishing a representative network of marine reserves,
comprising large-scale reserves on the high seas and a mosaic of smaller marine
reserves in association with well-managed, sustainable fishing areas within the
coastal zone. Plate 7.1 shows the locations of a global network of marine reserves

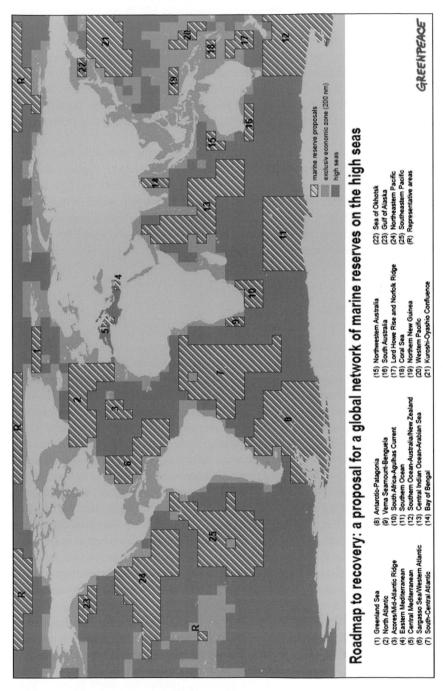

Plate 7.1 Proposed global network of marine reserves. The numbers on the map refer to locations of marine reserves that are listed in Appendix

for the high seas proposed by Greenpeace. The features of the selected marine reserves are given in Appendix. The following sections (up to Section 7.4.7) briefly describe the criteria that were used for the selection of marine reserve locations. For a fuller explanation, please refer to Roberts et al. (2006).

As a rule of thumb, marine reserves should be as large as possible given social constraints and should generally increase in size moving from near shore to offshore. Larger reserves are necessary in offshore regions because here the scale of animals' movements tends to be larger. Also, smaller reserves would be harder to identify in offshore areas, harder for fishers to comply with, and thus harder to enforce (Greenpeace 2006m; Roberts et al. 2006).

In planning a global network of marine reserves for the high seas, Greenpeace adopted a minimum reserve size of 5° latitude × 5° longitude. At the equator, this represents a size of approximately 560 × 560 km or 314,000 km^2 (Roberts et al. 2006).

In the coastal zone, building a network of smaller marine reserves will have the advantage of spreading fishery benefits to fishing communities along the coast, rather than concentrating them round a few large marine reserves with some communities losing their fishing grounds altogether.

7.4.3   Site Selection

Protected area selection must aim for broad representation of the full spectrum of biodiversity. It should not be driven, as it has been to date, simply by the need to protect threatened species or habitats. Selection based on the full spectrum of biodiversity and representative of the variation of habitats across the globe is necessary. Areas known to be important spawning and nursery grounds should be protected. With regard to migratory species, on land we are very familiar with the idea of using protected areas to safeguard highly migratory species. Dozens of migratory birds arrive and depart each year as they move between breeding and over-wintering habitats. We protect their breeding sites and their resting and refuelling spots along the way. This strategy can also be used in the sea. For nearly a century, fishery managers have protected the breeding sites of migratory species like herring and capelin when they gather in coastal shallows to spawn. They also protect juvenile nursery habitats to ensure that animals are able to grow undisturbed to marketable sizes (Roberts et al. 2006).

Many high-seas migratory species come close to coasts at some stage of their lives. Birds, turtles, seals and sea lions must come to land to breed. They can benefit from protected areas in national waters and on land.

There are also places on the high seas that support remarkable and predictable concentrations of life. Some of them are fixed, such as zones of high productivity around seamounts. Others are mobile, snaking around some general area of ocean at the whim of weather and currents or, like spinning eddies, may simply drift off then dissipate. These areas can be highly productive, such as upwellings where nutrient rich waters are brought to the surface, stimulating the growth of the plankton and, in turn, supporting other marine life. Convergence zones, where warm and cold currents press together and mix can also trigger dramatic plankton blooms and, similarly, they draw animals from far afield to feed or breed (Roberts et al. 2006).

Protecting mobile features on the high seas presents more of a challenge than protecting fixed locations like seamounts or fields of hydrothermal vents. In some places it may be feasible to create large enough marine reserves to incorporate most of the temporal variability in the location of these features. In others, this may not be possible. However, today's technologies provide the means by which we can both identify and protect important features as they move. Satellite sensors highlight areas of high chlorophyll concentration that indicate blooming plankton. Measures of sea surface height pick out currents and convergence zones, as do steep gradients in sea surface temperature.

Thermal data also show upwellings and coiling eddies. At present, satellite data are fed to fishing fleets either by national governments or private companies to guide boats to the places where it is known that fish will be concentrated. In the same way it would be a simple matter for governments to inform fishing fleets about that day, week or month's position of designated mobile high seas marine reserves (Norse et al. 2005). Mobile marine reserves could also be used to protect migrating species like turtles that follow predictable routes across the oceans. Fixed and mobile marine reserves would also benefit species such as birds that risk being killed in places where fishing operations are underway. In some places, marine reserves may promote the creation of oceanic hotspots of life, by allowing forage species to build in abundance and thereby attracting migratory animals to feed on them (Roberts et al. 2006).

The following are extracts from Roberts et al. (2006) and illustrate, by example, how sites were selected for the network of marine reserves proposed by Greenpeace (with full details provided in Appendix 1 of that report).

- "(1) We assembled maps of physical and biological data and brought them into a common format using the ArcInfo 3.3 ® Geographic Information System (GIS) program;
- (2) We created our own maps of use of the high seas by aquatic megafauna, drawing on data in numerous separate studies and reviews;
- (3) We consulted with experts in marine science and management, requesting them to nominate sites they believe should be afforded protection."

All the above sources of information were used to identify candidate sites for protection and to design a global network of high seas marine reserves. We offer a few comments on each approach below.

1. Mapping physical and biological data
We have brought together information from a wide variety of sources pertaining to the high seas. They include data showing species richness and species density of large pelagic fish species (billfish and tunas, Worm et al. 2005), sea surface temperature gradient (which identifies areas of mixing of cold and warm water masses), and the location of upwelling and downwelling areas, and bottom sediment types.

Fig. 7.1 Humpback whale breaching in waters off the coast of Tonga (Greenpeace/Paul Hilton)

We also obtained data on seamount distribution (www.seaaroundus.org), marine biomes and bathymetry. From the bathymetry data we calculated bottom complexity, a measure of habitat heterogeneity thought to relate to high benthic biodiversity, and have identified ocean trench habitats.

2. Air-breathing aquatic megafauna

Large and mobile animals seek out places on the high seas that are rich in their prey. Their movements can guide us to concentrations of marine life. We gathered and mapped data from dozens of studies of the movements of air-breathing aquatic megafauna, including albatrosses, pinnipeds (seals and sea lions), penguins and turtles. In particular, we sought data from studies using satellite, radio-tracking or data loggers to reveal high seas movements of these species. However, since there are insufficient data to be fully comprehensive, we concentrated on species for which there were good data and that spend significant amounts of their time on the high seas. We also developed maps showing species richness of cetaceans – whales, dolphins and porpoises (including the humpback whale, see Fig. 7.1) – across the world.

3. Expert consultation

We wrote to 404 experts on marine science and conservation asking them to nominate one or more candidate sites deserving high priority for protection on the high seas. We also requested they provided justification for their choice and send us supporting documentation, if any was available. We received replies from around 66. Experts nominated a wide range of sites, but there was also considerable consistency in their suggestions. In particular, seamounts and convergence zones between currents received multiple nominations for protection.

7.4.4   Networking and Connectivity of Marine Reserves

A protected area network, such as the global network proposed above, needs to be greater than the sum of its parts. A central objective for a network is to ensure eco-logical connectivity among protected area units. For species that move or disperse widely, populations in marine reserves should be mutually supporting. Levels of coverage (proposed as 40%), size (see above) replication and spacing (see below) of marine reserves need to be set taking connectivity considerations into account.

7.4.5   Level of Replication

Habitats should be replicated in more than one protected area in order to buffer against human or natural catastrophes that may damage or destroy populations in individual marine reserves. Furthermore, marine reserves cannot be mutually sup-porting unless there are similarities in the habitats and species they contain. The aims of replication are to spread the benefits of protection throughout the region, to provide insurance against human and natural impacts, and to ensure ecological connectivity among marine reserves.

How much replication is enough? At present this question has not been adequately resolved by science. Roberts and Hawkins (2000) suggested that all habitats should be replicated in at least three marine reserves. This level of replication was adopted in the recent re-zoning of the Great Barrier Reef Marine Park, and was applied within three biogeographic subdivisions of the park. However, when very large areas are involved, low levels of habitat replication in different marine reserves will fail to secure adequate connectivity. This is because the resulting marine reserves will tend to be very widely spaced and the distances between them may exceed the dispersal abilities of most resident species. Clearly, however, there are constraints on replica-tion. It may only be possible to attain low levels of replication for rare and isolated habitats. Higher levels of replication can be achieved by creating smaller marine reserves, but this could be self defeating if such areas are not sufficiently large to sustain populations of resident species (Roberts et al. 2006).

7.4.6   Spacing of Marine Reserves

Scales of ecological linkages in the high seas – i.e. the movement of juveniles and adult organisms, dispersal of their offspring, and transport of materials – extend from metres to thousands of kilometres. Seamount invertebrates, for example, may disperse only metres, while migratory tunas can undertake journeys of 20,000 km in a year. To ensure ecological connectivity in the network, marine reserves with similar habitats should generally be spaced from a few hundred to a few thousand kilometres apart.

7.4.7   Use of a Computer Program for Designing Networks

In their construction of a global map of high seas marine reserves, Greenpeace used a computer-assisted design procedure. *Marxan*, the computer program used, works by selecting sites for protection to create networks that meet user-defined conservation targets. It is the most widely used computer program for designing networks of marine reserves and has been instrumental in rezoning the Great Barrier Reef Marine Park in Australia (Roberts et al. 2006).

7.5   Implementation of Marine Reserves

At the international level, agreements which propose to establish marine protected areas include the Convention on Biological Diversity (CBD) and World Summit of Sustainable Development (WSSD). In 2004, parties to the CBD took a major step forward in committing to the establishment of a global network of marine protected areas by 2012 (Conference of Parties VII, Kuala Lumpar, Malaysia, February 2004). In the same vein, the WSSD Plan of Implementation included an agreement to establish marine protected areas by 2012 as a tool in the conservation and management of the oceans.

Other agreements on marine protected areas have been made at the regional level. For instance, in 1994, the Convention on the Protection of the Marine Environment of the Baltic Sea Area (the Helsinki Convention) set out its Recommendation 15/5 which states that "*the Contracting Parties shall take all appropriate measures to establish a system of coastal and marine Baltic Sea Protected Areas*". In 1998, the Convention for the Protection of the Marine Environment of the North-East Atlantic (the OSPAR Convention) agreed to protect and conserve biological diversity of the maritime area and its ecosystems, using the implementation of a network of marine protected areas as one measure to achieve this.

Although the Helsinki and OSPAR agreements provide a clear regional framework for establishing a network of marine protected areas in the North and Baltic Seas, there seems to be little political will to make such a network of marine protected areas a reality. Similarly, in the Mediterranean region, although there have been agreements to protect the Mediterranean Sea, agreements and commitments mean little without the political will to act. As a result, progress towards developing a network of marine reserves is still lacking. The development of a marine protection law in the European Union (EU), which covers a significant part of the North East Atlantic and parts of the Mediterranean and Black Sea, may, once adopted, bring new emphasis for the protection of those waters. As a minimum, it requires those states that are members of the EU to confirm their commitment to the above targets.

With regard to the high seas, there is currently no mechanism under the existing international framework provided by the United Nations Convention on the Law of the Sea (UNCLOS) and the Convention of Biodiversity (CBD) for implementing a global network of marine reserves. The CBD is the primary instrument providing direction to States for the establishment of marine protected areas and marine reserves under their jurisdiction, and also explicitly acknowledges the need for protective measures in areas beyond national jurisdiction. Article 4 of the Convention obliges Parties to apply the Convention to all processes or activities under their jurisdiction or control, including those taking place on the high seas. However, the Convention on Biological Diversity does not oblige States to take collective measures to protect the high seas and does not contain the necessary provisions to implement its 2012 goal of a comprehensive global network in areas beyond national jurisdiction.

It is Greenpeace's view that, in order to implement the CBD commitment and provide the necessary mandate to establish and manage marine reserves on the high seas, a new implementing agreement under UNCLOS is required. Such an implementing agreement would not require any amendment to the text of the Convention and would be consistent with article 22 (2) of the CBD which already obliges parties to implement the Convention *"with respect to the marine environment consistently with the rights and obligations of States under the Law of the Sea"*. The agreement would provide formal recognition of the need to protect biodiversity on the high seas, and a mandate to protect high seas areas for conservation purposes. Such an implementing agreement could be modelled on the UN Fish Stocks Agreement – which was itself negotiated in order to implement some of the Articles of UNCLOS – and could be used to address a number of gaps in the current governance of high seas biodiversity, in addition to those relating to the establishment of high seas marine reserves.

Such an approach is further justified by the fact that:

- UNCLOS is regarded as the framework agreement that delimits ocean areas and details State rights and duties in the high seas and the 'Area', and it is recognised as customary international law.
- UNCLOS' broad remit already covers most or all of the activities that impact on marine biodiversity, including emerging issues such as bioprospecting and noise pollution.
- UNCLOS provides a binding dispute settlement mechanism.

Such an agreement would build on, and provide for the implementation of, existing provisions in UNCLOS relating to the protection and preservation of the marine environment and the 'Area'.

Conclusion

This assessment of the state of the oceans bears witness to the continuing degradation of our seas at a local and global level. It highlights the scale of the impacts being inflicted on a wide range of marine habitats and species by unsustainable human activities, especially destructive fishing techniques and overfishing. Although the situation has deteriorated since the last review of this kind conducted by Greenpeace in 1998, there is also growing scientific evidence that the negative trends documented, in this report could be reversed. The implementation of the 'ecosystem approach' through the establishment of networks of large-scale, fully protected marine reserves and the application of sustainable management in the surrounding waters is the key to restoring the health of our oceans and maintaining the livelihoods of the many coastal communities dependent on them (Plate 1).

Plate 1 Octopus in an area identified by Greenpeace as a possible future marine reserve in the Mediterranean, near Menorca, Spain (Greenpeace/Gavin Newman)

There is still much to learn about the complex ecology of our oceans, but enough is known already for the world's governments and other ocean stakeholders to take positive action by creating marine reserves and ensuring that protection of the marine environment is at the core of all their marine policies and activities. We depend on the oceans for our continued existence, and so protecting the myriad of marine life – from the largest whales to the smallest planktonic organisms – is not only necessary for its own sake but for ours too, now and in the future.

Appendix

Areas Selected for a Global Network of Marine Reserves on the High Seas

For a map of the global network of marine reserves proposed by Greenpeace, see Plate 4.1.

This appendix lists the locations of the proposed marine reserves and gives some of their features (taken from Roberts et al. 2006).

Greenland Sea (1): This candidate marine reserve incorporates a swathe of the Greenland Sea between Svalbard and Eastern Greenland. It supports high levels of summer productivity and is an important feeding area for a variety of whales, seals and walrus, including hooded and harp seals (*Crystophoca cristata*) and (*Phagophilus groenlandicus*).

North Atlantic (2): This candidate marine reserve spans the North Atlantic from western Britain to Eastern Canada, including a southward extension west of Spain and Portugal. It includes a significant section of the mid-Atlantic ridge between 50°N and 60°N, incorporating the Charlie Gibbs Fracture Zone, making it important for deep-sea bottom life. It also includes the Rockall, Hatton and Porcupine Banks, deepwater areas to the west of the British Isles around 50°N–60°N. Further south, it incorporates the Josephine and Gorringe Banks off the coast of Spain. Josephine Bank is a current-swept seamount, rising from 3,200 m from the abyssal plain to within 170 m of the surface. It supports a diverse assemblage of fish, corals and other invertebrates and is a possible stepping-stone in tuna and turtle migrations. Deep-water trawling is seriously impacting this and other deepwater banks and seamounts that support important deepwater coral and sponge communities. Preliminary video surveys are showing that some banks still have pristine coral and other deep-water habitats but the fishing fleets are moving in fast and there is a high risk of these habitats being destroyed imminently.

The proposed reserve also holds the sub-polar front, a rich summer feeding area for many migratory fish, birds and mammals, as well as leatherback turtles (*Dermochelys coriacea*).

The reserve includes two shallow water regions off Eastern Canada, the 'tail' of the Grand Banks and Flemish Cap. The 'tail of the bank' is an important nursery area for cod (*Gadus morhua*) and other shallow water fish and currently supports a high-intensity international fishery with significant removals of undersized fish. Flemish Cap is a shallow water shelf covering a total area of 58,000 km² situated outside the Canadian EEZ. It supports the most heterogeneous offshore Atlantic cod population and is a foraging ground for northeastern and western Atlantic bluefin tuna populations (*Thunnus thynnus*). It is also visited by sperm whales (*Physeter macrocephalus*) among others and may be an important wintering area for great shearwaters (*Puffinus gravis*). Flemish Cap is currently intensively trawled.

Azores/Mid-Atlantic Ridge (3): This candidate reserve includes a section of the mid-Atlantic ridge that is rich in deep seamounts. It also covers the Rainbow Hydrothermal Vent Field which is located at 2,270–2,320 m depth. This comprises more than 30 groups of active small sulphide chimneys over an area of 15 km². About 32 different vent species have so far been recorded in the Rainbow area. The small spatial extent and site-specific communities make the vent field highly vulnerable to the increasing levels of scientific and commercial exploitation, including scientific sampling, bioprospecting and mining. Overlying waters of this region also support significant populations of whales and fish, including sperm whales.

Eastern Mediterranean (4): The area to the south, east and west of Cyprus is important for loggerhead (*Caretta caretta*) and green turtles (*Chelonia mydas*). The area to the south, extending to the Egyptian coast, contains the Eratosthenes seamount, an isolated and near-pristine mount with a likely highly endemic fauna. It also includes an area of biologically important deepwater cold-seeps off the Nile Delta.

Central Mediterranean (5): The Ligurian Sea between France, Northern Italy and Sardinia is already a multi-nationally managed marine protected area, the Pelagos reserve for marine mammals. The outstanding values and importance of this area for cetaceans was recognised in designating it as the first high seas marine protected area (MPA) in the Mediterranean. However, the cetaceans of the Ligurian Sea need to be protected from the impacts of fishing and other activities. To the immediate south, the Tyrrhenian Sea contains a significant concentration of seamounts, and has important migration routes for bluefin tuna (*Thunnus thynnus*), and is thought to be a major spawning area for the species in the Mediterranean.

Sargasso Sea/Western Atlantic (6): The Sargasso Sea lies to the west of the centre of the North Atlantic Gyre and is bounded on the west by the Gulf Stream. The Sargasso is a region of light winds and little rain. Coriolis forces acting on currents in the North Atlantic Gyre push water inward toward the centre of the gyre, and planetary rotation offsets it west. The Sargasso is thus a region of convergence of currents and gentle downwelling. It supports a high level of endemism among plankton species. Converging currents bring together flotsam and jetsam, and higher nutrients foster growth of seaweed mats that can cover huge areas. The western Sargasso and adjacent Gulf Stream is a hotspot for aquatic megafauna, including fish, turtles and marine mammals. Young juveniles of several species of turtle spend their time on the high seas in the Sargasso, feeding and sheltering among the seaweed mats. It is the breeding ground for threatened European eels (*Anguilla anguilla*) and is a migration route for whales, fish and turtles. Atlantic leatherback turtles migrate across the Sargasso from nesting beaches in Guyana to feeding grounds off Nova Scotia.

The southern spur of this candidate reserve also includes a section of the mid-Atlantic ridge that holds the Logatchev Hydrothermal Vent Area. Logatchev-1 field is characterised by three distinct sites: (i) a large sulphide mound with smoking craters; (ii) an active chimney complex called Irina-2; and (iii) an area with soft sediment and diffuse flow. These have a diversity of biotopes including thick bacterial mats, diffuse flow areas, and two different types of smokers – 'creeping' or horizontal smokers, and the more common vertical structures that resemble chimneys. Logatchev-1 also hosts an abundance of fauna, including swarms of shrimp at black smokers, clam beds in the sediment biotopes, mussels from the genus *Bathymodiolus* on sulphide chimneys and sulphide base areas, as well as sea anemones. Logatchev-2 has six sulphide mounds and extensive massive sulphide deposits containing high concentrations of copper, gold, zinc, uranite (uranium), and the highest concentration of cobalt of any hydrothermal vent field recorded to date. The main present threat is scientific research; future threats are mining and bioprospecting.

South-Central Atlantic (7): This proposed reserve includes a large section of the mid-Atlantic ridge that is rich in seamounts, as well as a ridge spur that extends from the coast of Brazil into the Atlantic, the Trindade ridge. This ridge extends west to east at a latitude centred on 20°S, terminating at the [small Brazilian] islands of Trindade and Martin Vas. It has high endemism of species such as fish and is currently being damaged by deep sea trawling. The proposed reserve also includes a critical migration route for green turtles at approximately 5°S–10°S, between breeding beaches in Ascension and feeding grounds off South America. The candidate reserve also supports exceptional species diversity of tuna, billfish and toothed whales. The far southern part of the region includes a transition area between South Atlantic and Southern Ocean waters and is a critical feeding area for albatrosses, penguins and pinnipeds that breed on remote islands of the South Atlantic.

Antarctic-Patagonia (8): This proposed reserve incorporates the Antarctic Peninsula, the entire Weddell Sea and Bellingshausen Seas, much of the Scotia Sea, and parts of the South Atlantic. It includes the Patagonian Shelf edge, a region of exceptionally high productivity. This is the convergence zone between the south-flowing warm Brazil current and the northward-flowing cold water Malvinas-Falklands current. The proposed reserve incorporates transitional waters between the warmer southern Atlantic and cold polar seas, which also support high productivity. It is a key area for aquatic megafauna, including albatross, seals, penguins and whales. It supports 8 species of baleen whales and a further 20 species of toothed whale and dolphin. There is an important deep-sea fishery for Patagonian toothfish (*Dissostichus eleginoides*) and a highly productive shallower water fishery for squid, both of which have serious impacts on wildlife.

Vema Seamount-Benguela (9): This proposed reserve includes Vema Seamount which lies approximately 500 km off the coasts of Namibia and South Africa. It is an isolated seamount, rising from a deep abyssal plain that supports a high diversity of fish and invertebrates, probably with significant endemism. It has suffered badly from over-exploitation of rock lobsters. The reserve also includes part of the Benguela current ecosystem which sustains highly productive fisheries for pelagic and bottom living species, but increasingly, deepwater species like those on the seamount are being targeted.

South Africa-Agulhas Current (10): This proposed reserve incorporates the confluence between the warm, south-flowing Agulhas Current and cool, nutrient-rich waters from the Southern Ocean. The mixing area includes large warm core eddies that propagate eastwards from the tip of South Africa, promoting high productivity throughout the area. This productivity sustains many resident and migratory species, including southern right whales (*Eubalaena glacialis*), sharks, billfish, penguins and albatross. The proposed reserve also includes migration routes for leatherback turtles (*Dermochelys coriacea*) that nest on the South African and Mozambique coasts.

Southern Ocean (11): This proposed reserve incorporates waters between approximately 45°S to the edge of Antarctica. These waters are dotted with small islands that are critical breeding sites for seabirds and marine mammals, including the likes of Prince Edward, Crozet, and Kerguelen. The region is important for air-breathing megafauna. Hotspots of use of the ocean by these species are concentrated around several of these sub-Antarctic islands. The region also has a complex of seamounts including the Afrikaner II Rise (ca. 46°S, 42°E), which straddles the edge of the South African EEZ boundary. This is the site of a large amount of illegal fishing activity. It is also a favoured foraging site for many top predators, especially albatrosses, breeding both on the Prince Edward Islands and the Crozets. The area also contains the isolated Ob and Lena Tablemounts.

Southern Ocean-Australia/New Zealand (12): This region of the Southern Ocean is critical for wildlife. It includes the area between Australia, New Zealand and Antarctica, centred on Macquarie Island. This incorporates Campbell Islands, The Antipodes Islands and Balleny Islands as well as the western half of the Ross Sea. In addition, the proposed reserve covers a large area of seamounts that are subject to bottom fishing, including the Macquarie Ridge and South Tasman Rise. Collectively the islands, several hundred seamounts and the comparatively shallow surrounding seas represent a strong physical barrier in oceanographic terms. The marine environment is thus one of a slow-moving water mass rich in nutrients. The region shares similar features of high productivity and high concentration of air-breathing megafauna as the Southern Ocean Reserve, described above. It supports five species of penguin, for example, of which two are endemic, the snares (*Eudyptes robustus*) and royal (*Eudyptes schegel*). It is also important for New Zealand sea lions (*Phocarctos hookeri*) and has a rich but poorly described underwater biota, including fragile seamount communities. The Ross Sea is distinct from the wider Antarctic marine ecosystem. Its highly productive and healthy food web includes such charismatic megafauna as whales, seals and penguins, but it is imminently threatened by the rapid growth of toothfish fishing and hunting of minke whales (*Balaenoptera acutorostrata*). Potential threats may arise from bioprospecting, tourism and the introduction of invasive marine species from ship hulls.

Central Indian Ocean-Arabian Sea (13): This region incorporates the central-western Indian Ocean. It includes the Saya de Malha Banks, a part of the Mascarene Plateau between the Seychelles, Madagascar, Mauritius and Chagos. The proposed reserve supports deep and shallow water fauna, including diverse coral reefs. The area contains the largest coral reef and seagrass habitat in the world in international waters, and is a stepping stone, providing connectivity across the entire Indian Ocean. As such it is crucial to gene flow and migratory stocks. The area around Saya de Malha is a major whale calving ground. The proposed reserve incorporates a region of high tuna and billfish diversity as well as major seamount areas such as the Ninetyeast Ridge, Broken Ridge and the seafloor-spreading zone of the Mid-Indian Ridge. Much of the proposed reserve is subject to high fishing intensities including trawling, longlining and purse-seining. Distant water fleets from many nations, such as those of Europe, Sri Lanka and Asia, go there to fish. The Chagos Islands, which the reserve bounds, are uninhabited and almost unpolluted and little affected by direct human impacts except fishing. The proposed reserve surrounds these and islands of the Maldives and Lakshadweep groups, providing an offshore conservation buffer. The western part of the reserve extends north into the Arabian Sea, encompassing a region of seasonal upwelling driven by monsoon winds.

Bay of Bengal (14): This region supports a high diversity of large pelagic fishes, including tunas, billfish and whale sharks. It is an important migration route for turtles breeding on India's Orissa coast and is intensively fished.

Northwestern Australia (15): This proposed reserve incorporates a region of exceptional richness of tunas and billfishes. It is a spawning area for southern bluefin tuna (*Thunnus maccoyi*i). It is also a critical feeding and juvenile habitat for highly migratory whale sharks.

South Australia (16): This region supports very high productivity and is the meeting place of the warm, south-flowing coastal Leeuwin Current and the cool water of the Southern Ocean. The Leeuwin spills around Cape Leeuwin into the Great Australian Bight, mixing as it does with cooler, nutrient-rich water, fuelling plankton growth. This mixing zone attracts large concentrations of aquatic megafauna to feed. It is home to over fifteen species of toothed whale and dolphin, and eight species of baleen whale.

Lord Howe Rise and Norfolk Ridge (17): This area supports extensive seamount chains of the Lord Howe Rise and Norfolk Ridge. Recent scientific studies indicate they support a highly endemic fauna, with a quarter to a third of species sampled found nowhere else. Many appear to be relict species. Deepwater bottom trawlers intensively exploit the seamounts. This region supports a high diversity of large pelagic fishes.

Coral Sea (18): This proposed reserve incorporates a region of exceptional richness of tunas, billfishes and other large pelagic animals, as well as being a global centre of endemism for coral reef fauna. It contains significant deepwater seamount and slope habitats, also supporting high levels of endemism.

Northern New Guinea (19): This area represents an important migration corridor for critically endangered Pacific leatherback turtles. It also incorporates the Eauripik Rise, a significant seamount area in one of the most biologically diverse regions of the world's oceans.

Western Pacific (20): This area is bounded by the Exclusive Economic Zones of the Federated States of Micronesia, Papua New Guinea, Solomon Islands, Tuvalu, Kiribati, Nauru and Marshall Islands. It represents an important spawning area for tunas, including bigeye tuna (*Thunnus obesus*).

Kuroshi-Oyashio Confluence (21): This proposed reserve covers the area of convergence between the warm north-flowing Kuroshio Current and the cold south-flowing Oyashio Current. It is a highly productive region that is rich in fish and aquatic megafauna, including whales, dolphins, tuna and albatross. These seas are a productive fishing ground for sardines, squid, bonito, and mackerel. In deep water, the proposed reserve includes sections of the Japan and Kuril Trenches.

Sea of Okhotsk (22): The Sea of Okhotsk is a highly productive semi-enclosed sea that supports large populations of fish and shellfish, together with high concentrations of marine mega-fauna, including the critically endangered western population of Pacific gray whales. It is an important bird and fish migration route, especially for salmon spawning in rivers of eastern Russia. In winter, large areas of the sea are ice-bound, making for highly seasonal productivity. Overfishing and illegal fishing are widespread, and the region is being rapidly developed for oil production, threatening wildlife.

Gulf of Alaska (23): This region in the Gulf of Alaska is highly productive and is a key area for migrating salmon, salmon sharks, whales, for example, gray whale (*Eschrichtius robustus*), fur seals (*Callorhinus ursinus*) and Steller's sea lions (*Eumetopias jubatus*), among others. It also contains a significant concentration of seamounts.

Northeastern Pacific (24): This large equatorial/sub-equatorial area extends from the southwest towards the North American coast. It supports high pelagic fish diversities and abundance and lies on a major trans-Pacific pathway for tuna migration. It also contains key migration corridors for leatherback turtles (*Dermochelys coriacea*) across the Pacific from New Guinea, and from nesting beaches in California and Mexico to feeding grounds that the proposed reserve largely includes. It also is important for loggerhead turtles (*Caretta caretta*), blackfooted albatross (*Phoebastria nigripes*) and northern elephant seals (*Mirounga angustirotris*). In deep water, the region includes the Clarion-Clipperton Fracture Zone, an area with high concentrations of manganese nodules. These nodules support a hard substrate fauna on an otherwise sediment-covered abyssal plain. Without protection they are likely to be impacted by future mining operations.

Southeastern Pacific (25): This proposed reserve incorporates the region bounded by Galapagos-Cocos-Panama-Costa Rica-Columbia-Ecuador. There is considerable potential for cooperative protection of high seas areas adjacent to the EEZs of these countries and an initiative is already underway, coordinated by Conservation International, to secure such cooperation. The Humboldt Current bounds this region to the east before flowing away from the Peruvian coast toward the Galapagos. Consequently this is an area of high aquatic productivity, but it is also a convergence between very different bodies of water, including North and South Equatorial Currents. The area is intensively fished but it supports a diverse and still abundant megafauna. It includes a migration route for the critically endangered Pacific leatherback turtles (*Dermochelys coriacea*) from breeding beaches in Costa Rica to feeding grounds, which the proposed reserve largely encompasses. To the southwest, the region is a hotspot of tuna and billfish diversity. The region also contains important and vulnerable deepwater habitats, including the Galapagos Rise, Sala y Gomez Ridge and Challenger Fracture Zone.

Representative Areas (Marked 'R'): The global network of marine reserves presented here has been designed to be fully representative of high seas habitats and biodiversity. Some of the areas included are not described separately here, and are simply marked 'R'. These areas include important representatives of particular ecosystems, such as bottom types, depth zones and biogeographic regions, needed to achieve targets of coverage and representation of habitats.

References

Aarkrog, A., Dahlgard, H. and Nielsen, S.P. (2000). Environmental radioactive contamination in Greenland: a 35 years retrospect. The Science of the Total Environment 245: 233–248.

ACIA (2004). Impacts of a Warming Arctic: Arctic Climate Impact Assessment. Cambridge University Press, Cambridge. 140 pp.

Adey, W.H., McConnaughey, T.A., Small, A.M. and Spoon, D.M. (2000). Coral reefs: endangered, biodiverse, genetic resources. In: Seas at the Millennium: An Environmental Evaluation. Volume III, Global Issues and Processes (eds. C. Sheppard), Ch. 109, pp. 33–42. Pergamon/Elsevier Science, Oxford. ISBN: 0-08-043207-7.

Adinehzadeh, M., Reo, N.V., Jarnot, B.M., Taylor, C.A. and Mattie, D.R. (1999). Dose-response hepatotoxicity of the peroxisome proliferator, perfluorodecanoic acid and the relationship to phospholipid metabolism in rats. Toxicology 134: 179–195.

Alaee, M., Arias, P., Sjödin, A. and Bergman, A. (2003). An overview of commercially used brominated flame retardants, their applications, their use patterns in different countries/regions and possible modes of release. Environment International 29: 683–689.

Albaigés, J. (2006). The Prestige oil spill: a scientific response. Marine Pollution Bulletin 53 (5–7): 205–207.

Allain, M. (2007). Trading away our oceans: why trade liberalization of fisheries must be abandoned. Greenpeace International, Amsterdam, The Netherlands. 75 pp.

Alley, R.B., Clark, P.U., Hybrechts, P. and Joughin, I. (2005). Ice sheet and sea-level changes. Science 310: 456–461.

Allsopp, M., Santillo, D., Johnston, P. and Stringer, R. (1999) The Tip of the Iceberg?: State of Knowledge on Persistent Organic Pollutants in Europe and the Arctic. Publ. Greenpeace International, August 1999, ISBN: 90-73361-53-2: 76 pp.

Allsopp, M., Santillo, D., Walters, A. and Johnston, P. (2005). Perfluorinated chemicals: an emerging concern. Greenpeace Research Laboratories, Technical Note 04/2005.

Allsopp, M., Walters, A., Santillo, D. and Johnston, P. (2006). Plastic Debris in the World's Oceans. Greenpeace International, Amsterdam, The Netherlands. 44 pp. [http://oceans.greenpeace.org/raw/content/en/documents-reports/plastic_ocean_report.pdf].

Allsopp, M., Johnston, P. and Santillo, D. (2008). Challenging the aquaculture industry on sustainability. Greenpeace Research Laboratories Technical Note 01/2008. [http://www.greenpeace.to/publications/Aquaculture_Report_Technical.pdf].

Alverson, D.L., Freeberg, M.H., Murawski, S.A. and Pope, J.G. A global assessment of fisheries bycatch and discards. FAO Fisheries Technical Paper No. 339. Food and Agricultural Organization of the United Nations, Rome, 1994. 233p.

Amaya, E., Davis, D.A. and Rouse, D.B. (2007). Alternative diets for the Pacific white shrimp Litopenaeus vannamei. Aquaculture 262: 419–425.

American Association for the Advancement of Science (2001). Scientific Consensus Statement on marine reserves and marine protected areas. Annual Meeting of the American Association for the Advancement of Science, 17 February 2001.

Anderson, L. (2004). Genetically engineered fish – new threats to the environment. Greenpeace International, Amsterdam, The Netherlands. 20 pp.

Anker-Nilssen, T., Barrett, R.T. and Krasnov, J.K. (1997). Long- and short-term responses of seabirds in the Norwegian and Barents Seas to changes in stocks of prey fish. Forage Fishes in Marine Ecosystems. Proceedings of the International Symposium on the Role of Forage Fishes in Marine Ecosystems. University of Alaska, Fairbanks, AK, pp. 683–698.

Arnold, J.M., Brault, S. and Croxall, J.P. (2006). Albatross populations in peril: a population trajectory for black-browed albatrosses at South Georgia. Ecological Applications 16(1): 419–432.

Asplund, L., Bignert, A. and Nylund, K. (2004). Comparison of spatial and temporal trends of methoxylated PBDEs, PBDEs, and hexabromocyclododecane in herring along the Swedish coast. Organohalogen Compounds 66: 3988–3993.

Atkinson, A., Siegel, V., Pakhomov, E. and Rothery, P. (2004). Long-term decline in krill stock and increase in salps within the Southern Ocean. Nature 432: 100–103.

Austin, M.E., Kasturi, B.S., Barber, M., Kannan, K., MohanKumar, P.S. and MohanKumar, S.M.J. (2003). Neuroendocrine effects of perfluorooctane sulphonate in rats. Environmental Health Perspectives 111(12): 1485–1489.

Australian Institute of Marine Science and Global Coral Reef Monitoring Network (2000). Status of Coral Reefs of the World 2000 (ed. C. Wilkinson), 363 pp. Australian Insititute of Marine Science, Cape Ferguson, Queensland, Australia. ISBN: 0-642-32209-0.

Australian Institute of Marine Science and Global Coral Reef Monitoring Network (2004). Status of Coral Reefs of the World 2004. Volume 1 (ed. C. Wilkinson), 301 pp. Australian Institute of Marine Science, Australia. ISSN 1447–6185.

Australian Marine Conservation Society (2008). Farming Australia's seas. It no solution. Accessed January 2008 at: http://www.amcs.org.au/default2.asp?active_page_id=159

Avnimelech, Y. (2006). Biofilters: the need for a new comprehensive approach. Agricultural Engineering 34: 172–178.

Badalamenti, F., D'Anna, G., Pinnegar, J.K. and Polunin, N.V.C. (2002). Size-related trophodynamic changes in three target fish species recovering from intensive trawling. Marine Biology 141: 561–570.

Baker, A.C., Starger, C.J., McClanahan, T.R. and Glynn, P.W. (2004). Coral's adaptive response to climate change. Nature 430: 741.

Baker, G.B. and Wise, B.S. (2005). The impact of pelagic longline fishing on the flesh-footed shearwater Puffinus carneipes in Eastern Australia. Biological Conservation 126: 306–316.

Baker, J.D., Littnan, C.L. and Johnston, D.W. (2006). Potential effects of sea level rise on the terrestrial habitats of endangered and endemic megafauna in the Northwestern Hawaiian Islands. Endangered Species Research 4: 1–10.

Baran, E. and Hambrey, J. (1998). Mangrove conservation and coastal management in Southeast Asia: what impact on fishery resources? Marine Pollution Bulletin 37(8–12): 431–440.

Barnes, K.N., Ryan, P.G. and Boix-Hinzen, C. (1997). The impact of the hake Merluccius spp. longline fishery off South Africa on Procellariiform seabirds. Biological Conservation 82: 227–234.

Barraclough, S. and Finger-Stich, A. (1996). Some ecological and social implications of commercial shrimp farming in Asia. United Nations Research Institute for Social Development, Geneva, Switzerland.

Barrett, G., Caniggia, M.I. and Read, L. (2002). "There are more vets than doctors in Chiloé": social and community impact of the globalization of aquaculture in Chile. World Development 30(11): 1951–1965.

Barton, A.D. and Casey, K.S. (2005). Climatological context for large-scale coral bleaching. Coral Reefs 24: 536–554.

Baum, J.K., Myers, R.A., Kehler, D.G., Worm, B., Harley, S.J. and Doherty, P.A. (2003). Collapse and conservation of shark populations in the Northwest Atlantic. Science 299: 389–392.

Beardmore, J.A., Mair, G.C. and Lewis, R.I. (1997). Biodiversity in aquatic systems in relation to aquaculture. Aquaculture Research 28: 829–839.

Bearzi, G., Reeves, R.R., Notarbartolo-di-sciara, G., Politi, E., Cañadas, A., Frantzis, A. and Mussi, B. (2003). Ecology, status and conservation of short-beaked common dolphins *Delphinus delphis* in the Mediterranean Sea. Mammal Review 33(3): 224–252.

Beaugrand, G., Brander, K.M., Lindley, J.A., Souissi, S. and Reid, P.C. (2003). Plankton effect on cod recruitment in the North Sea. Nature 426: 661–664.

Beck, M.W., Heck, K.L., Able, K.W., Childers, D.L., Eggleston, D.B., Gillanders, B.M., Halpern, B., Hays, C.G., Hoshino, K., Minello, T.J., Orth, R.J., Sheridan, P.F. and Weinstein, M.P. (2001). The identification, conservation, and management of estuarine and marine nurseries for fish and invertebrates. Bioscience 51(8): 633–641.

Beger, M., Jones G.P. and Munday P.L. (2003). Conservation of coral reef biodiversity: a comparison of reserve selection procedures for corals and fishes. Biological Conservation 111: 53–62.

Belda, E.J. and Sánchez, A. (2001). Seabird mortality on longline fisheries in the western Mediterranean: factors affecting by-catch and proposed mitigating measures. Biological Conservation 98: 357–363.

Bellwood, D.R., Hughes, T.P., Folke, C. and Nyström, M. (2004). Confronting the coral reef crisis. Nature 429: 827–833.

Bergman, M.J.N. and van Santbrink, J.W. (2000). Mortality in megafaunal benthic populations caused by trawl fisheries on the Dutch continental shelf in the North Sea in 1994. ICES Journal of Marine Science 57: 1321–1331.

Berthiaume, J. and Wallace, K.B. (2002). Perfluorooctanoate, perfluorooctanesulfonate, and N-ethyl perfluorooctanesulfonamido ethanol; peroxisome proliferation and mitochondrial biogenesis. Toxicology Letters 129(1–2): 23–32.

Beveridge, M.C.M., Ross, L.G. and Stewart, J.A. (1997). The development of mariculture and its implications for biodiversity. In: Marine Biodiversity: Patterns and Processes (eds. R.F.G. Ormond, J.D. Gage and M.V. Angel), Ch. 16, pp. 372–393. Cambridge University Press, Cambridge.

Birdlife International (2005). Fisheries organisations failing to safeguard the world's albatrosses. Accessed 2007 at: http://www.birdlife.org/news/pr/2005/03/rfmos.html

Birkeland, C. (1997). Introduction. In: Life and Death of Coral Reefs (ed. C. Birkeland), Ch. 1, pp. 1–12. Chapman & Hall, New York.

Birnbaum, L.S. and Staskal, D.F. (2004). Brominated flame retardants: cause for concern? Environmental Health Perspectives 112(1): 9–17.

Biscoito, M., Segonzac, M., Almeida, A.J., Desbruyères, D., Geistdoerfer, P., Turnipseed, M. and Van Dover, C. (2002). Fishes from the hydrothermal vents and cold seeps – an update. Cahiers de Biologie Marine 43: 359–362.

Block, B.A., Teo, S.L.H., Walli, A., Boustany, A., Stokesbury, M.J.W., Farwell, C.J., Weng, K.C., Dewar, H. and Williams, T.D. (2005). Electronic tagging and population structure of Atlantic bluefin tuna. Nature 434: 1121–1126.

Bodkin, J.L., Ballachey, B.E., Dean, T.A., Fukuyama, A.K., Jewett, S.C., McDonald, L., Monson, D.H., O'Clair, C.E. and VanBlaricom, G.R. (2002). Sea otter population status and the process of recovery from the 1989 'Exxon Valdez' oil spill. Marine Ecology Progress Series 241: 237–253.

Bonsdorff, E., Rönnberg, C. and Aarnio, K. (2002). Some ecological properties in relation to eutrophication in the Baltic Sea. Hydrobiologia 475–476(1): 371–377.

Borges, L., Rogan, E. and Officer, R. (2005). Discarding by the demersal fishery in the waters around Ireland. Fisheries Research 76: 1–13.

Bossi, R., Riget, F.F. and Dietz, R. (2005). Temporal and spatial trends of perfluorinated compounds in ringed seal (*Phoca hispida*) from Greenland. Environmental Science and Technology 39(19): 7416–7422.

Bours, H. and Losada, S. (2007). Witnessing the Plunder 2006. How Illegal Fish from West African Waters Finds Its Way to the EU Ports and Markets. Greenpeace International, Amsterdam, The Netherlands. 40 pp.

Bours, H., Gianni, M. and Mather, D. (2001). Pirate Fishing Plundering the Oceans. Greenpeace International, Amsterdam, The Netherlands. 29 pp. [http://archive.greenpeace.org/oceans/reports/pirateen.pdf].

Boyd, C.E. (2002). Mangroves and coastal aquaculture. In: Responsible Marine Aquaculture (eds. R.R. Stickney and J.P McVey), Ch. 9, pp. 145–158. CABI, Wallingford, Oxon. ISBN: 0-85199-604-3.

Brander, K. (1981). Disappearance of common skate *Raia batis* from Irish Sea. Nature 290: 48–49.

Braune, B.M. and Simon, M. (2004). Trace elements and halogenated organic compounds in Canadian Arctic seabirds. Marine Pollution Bulletin 48: 986–1008.

Brigden, K., Labunska, I., Santillo, D. and Allsopp, M. (2005). Recycling of electronic wastes in China and India: workplace and environmental contamination. Greenpeace Research Laboratories Technical Note 09/2005. Greenpeace International, Amsterdam, The Netherlands. 56 pp. (+ 47 pp. appendices). [http://www.greenpeace.to/].

Briggs, J.C. (2005). Coral reefs: conserving the evolutionary sources. Biological Conservation 126: 297–305.

Broecker, W.S. (1997). Thermohaline circulation, the Achilles heel of our climate system: will man-made CO_2 upset the current balance? Science 278(5343): 1582–1588.

Brothers, N.P., Cooper, J. and Løkkeborg, S. (1999). The incidental catch of seabirds by longline fisheries: worldwide review and technical guidelines for mitigation. Food and Agricultural Organisation of the United Nations, Rome. FAO Fisheries Circular No. 937. 101 pp.

Browdy, C., Seaborn, G., Atwood, H., Davis, D.A., Bullis, R.A., Samocha, T.M., Wirth, E. and Leffler, J.W. (2006). Comparison of pond production efficiency, fatty acid profiles, and contaminants in *Litopenaeus vannamei* fed organic plant-based and fish-meal-based diets. Journal of the Aquaculture Society 37(4): 437–451.

Brown, B.E. (1997). Disturbances to reefs in recent times. In: Life and Death of Coral Reefs (eds. C. Birkeland), Ch. 15, pp. 354–379. Chapman & Hall, New York.

Bruno, J.F., Petes, L.E., Harvell, C.D. and Hettinger, A. (2003). Nutrient enrichment can increase the severity of coral diseases. Ecology Letters 6: 1056–1061.

Buchmann, K., Uldal, A. and Lyholt, H.C.K. (1995). Parasitic infections in Danish trout farms. Acta Vetinaria Scandanavia 36(3): 283–298.

Burreau, S., Zebühr, Y., Ishaq, R. and Broman, D. (2000). Comparison of biomagnification of PBDEs in food chains from the Baltic Sea and the Northern Atlantic Sea. Organohalogen Compounds 47: 253–255.

Buschmann, A.H., Riquelme, V.A., Hernández-Gonález, D., Varela, D., Jiménez, J.E., Henríquez, L.A., Vergara, P.A., Guíñez, R. and Filún, L. (2006). A review of the impacts of salmonid farming on marine coastal ecosystems in the southeast Pacific. ICES Journal of Marine Science 63: 1338–1345.

Bytingsvik, J., Gaustad, H., Salmar, M.P., Soermo, E.G., Baek, K., Føreid, S., Ruus, A., Skaare, J.U. and Jenssen, B.M. (2004). Spatial and temporal trends of BFRs in Atlantic cod and Polar cod in the North-East Atlantic. Organohalogen Compounds 66: 3918–3922.

California Department of Fish and Game (2007). Calfornia Department of Fish and Game, Marine Region. Marine Life Protection Act Initiative. Accessed 2007 at: http://www.dfg.ca.gov/mrd/mlpa/ccmpas.html

Cardinale, M. and Svedäng, H. (2004). Modelling recruitment and abundance of Atlantic cod, *Gadus morhua*, in eastern Skagerrak-Kattegat (North Sea): evidence of severe depletion due to a prolonged period of high fishing pressure. Fisheries Research 69: 263–282.

Carreras, C., Cardona, L. and Aguilar, A. (2004). Incidental catch of the loggerhead turtle *Caretta caretta* off the Balearic Islands (western Mediterranean). Biological Conservation 117: 321–329.

Casale, P., Laurant, L. and De Metrio, G. (2004). Incidental capture of marine turtles by the Italian trawl fishery in the north Adriatic Sea. Biological Conservation 119: 287–295.

Casey, J.M. and Myers, R.A. (1998). Near extinction of a large widely distributed fish. Science 281: 690–692.

Caswell, H., Brault, S., Read, A.J. and Smith, T.D. (1998). Harbor porpoise and fisheries: an uncertainty analysis of incidental mortality. Ecological Applications 8(4): 1226–1238.

Caswell, H., Fujiwara, M. and Brault, S. (1999). Declining survival probability threatens the North Atlantic right whale. Proceedings of the National Academy of Sciences of the United States of America 96(6): 3308–3313.

CBRC (2006). Cetacean by-catch facts. Cetacean By-Catch Resource Centre. Accessed 2006 at: http://www.cetaceanbycatch.org/status.cfm

CCC/STSL (2003). Sea turtles threats and conservation. Caribbean Conservation Corporation & Sea Turtle Survival League. Accessed 2006 at: http://www.cccturtle.org/sea-turtle-information. php?page=threats

Chopin, T. (2006). Integrated multi-trophic aquaculture. What is it and why you should care... and don't confuse it with "polyculture". Northern Aquaculture 12(4): 4 (July 2006).

Christensen, V., Guénette, S., Heymans, J.J., Walters, C.J., Watson, R., Zeller, D., and Pauly, D. (2003). Hundred-year decline of North Atlantic predatory fishes. Fish and Fisheries 4: 1–24.

Clapham, P.J., Young, S.B. and Brownell, R.L. (1999). Baleen whales: conservation issues and the status of the most endangered populations. Mammal Review 29(1): 35–60.

Clapham, P.J., Childerhouse, S., Gales, N.J., Rojas-Bracho, L., Tillman, M.F. and Brownell, R.L. (2007). The whaling issue: conservation, confusion, and casuistry. Marine Policy 31: 314–319.

Clark, M. (2001). Are deepwater fisheries sustainable? – the example of orange roughy (Hoplostethus atlanticus) in New Zealand. Fisheries Research 51: 123–135.

Clark, M. and O'Driscoll, R. (2003). Deepwater fisheries and aspects of their impact on seamount habitat in New Zealand. Journal of the Northwest Atlantic Fishery Science 31: 441–458.

Cobbing, M. (2008). Toxic Tech: Not in Our Backyard. Uncovering the Hidden Flows of E-Waste. Greenpeace International, Amsterdam, The Netherlands. 76 pp.

Cockerham, L.G. and Cockerham, M.B. (1994). Environmental ionising radiation. In: Basic Environmental Toxicology (eds. L.G. Cockerham and B.S. Shane), Ch. 9, pp. 231–261. CRC, Boca Raton, FL.

Commission of the European Communities (2005). Proposal for a Directive of the European Parliament and of the Council relating to restrictions on the marketing and use of perflurooctane sulfonates (amendment of Council Directive 76/769/EEC). Brussels, 5 December 2005.

Commonwealth of Australia (2003). The benefits of marine protected areas. A discussion paper prepared for the Vth IUCN World Parks Congress Durban, South Africa 2003. ISBN: 0-624-54949-4.

Conquest, L. and Tacon, A. (2006). Utilization of microbial floc in aquaculture systems: a review. Meeting Abstract. Presented at: World Aquaculture Society, America Meeting, Las Vegas, NV. Microbial Controlled Systems. Special symposium, 15 February 2006.

Conservation International (2008). Bird's head. Accessed February 2008 at: http://www.conservation.org/explore/priority_areas/oceans/Pages/birdshead.aspx

Cousins, K.L., Dalzell, P. and Gilman, E. (2000). Appendix 1. Managing pelagic longline – albatross interactions in the North Pacific Ocean. In: Albatross and Petrel Mortality from Longline Fishing (ed. J. Cooper). International workshop, Honolulu, HI11–12 May 2000. Report and Presented Papers. Marine Ornithology 28: 159–174.

Covaci, A., Gerecke, A.C., Law, R.J., Voorspoels, S., Kohler, M., Heeb, N.V., Leslie, H., Allchin, C.R. and de Boer, J. (2006). Hexabromocyclodecanes (HBCDs) in the environment and humans: a review. Environmental Science and Technology 40(12): 3679–3688.

Cuthbert, R., Hilton, G., Ryan, P. and Tuck, G.N. (2005). At sea distribution of breeding Tristan albatrosses Diomedea dabbenena and potential interactions with pelagic longline fishing in the South Atlantic Ocean. Biological Conservation 121: 345–355.

D'Agrosa, C., Lennert-Cody, C.E. and Vidal, O. (2000). Vaquita by-catch in Mexico's artisanal gillnet fisheries: driving a small population to extinction. Conservation Biology 14(4): 1110–1119.

Dans, S.L., Alonso, M.K., Pedraza, S.N. and Crespo, E.A. (2003). Incidental catch of dolphins in trawling fisheries off Patagonia, Argentina: can populations persist? Ecological Applications 13(3): 754–762.

Darnerud, P.O. (2003). Toxic effects of brominated flame retardants in man and wildlife. Environment International 29: 841–853.

Das, B., Khan, Y.S.A. and Das, P. (2004). Environmental impact of aquaculture-sedimentation and nutrient loadings from shrimp culture of the southeast coastal region of the Bay of Bengal. Journal of Environmental Sciences 16(3): 466–470.

Deep Sea Conservation Coalition (2005). Debunking claims of sustainability: high seas bottom trawl red herrings. Prepared for the DSCC by the Marine Conservation Biology Institute. DSCC, April 2005. 16 pp.

DEFRA (2005a). Consultation on Proposed National Action to Restrict the Use of Perfluorooctane Sulphonate (PFOS) and Substances That Degrade to PFOS, Annex A – Draft Regulations. Department for Environmental Food and Rural Affairs. 5 pp.

DEFRA (2005b) e-Digest of Environmental Statistics: Coastal and Marine Waters, Department for Environment, Food and Rural Affairs, December 2005.

Derocher, A.E., Lunn, N.J. and Stirling, I. (2004). Polar bears in a warming climate. Integrative and Comparative Biology 44: 163–176.

Deutsch, L., Graslund, S., Folke, C., Troell, M., Huitric, M., Kautsky, N. and Lebel, L. (2007) Feeding aquaculture growth through globalization: Exploitation of marine ecosystems for fishmeal. Global Environmental Change 17: 238–249

Devine, J.A., Baker, K.D and Haedrich, R.L. (2006). Deep-sea fishes qualify as endangered. Nature 439: 29.

De Wit, C.A. (2002). An overview of brominated flame retardants in the environment. Chemosphere 46: 583–624.

De Wit, C.A., Alaee, M. and Muir, D.C.G. (2006). Levels and trends of brominated flame retardants in the Arctic. Chemosphere 64(2): 209–233.

Diaz, R.J. (2001). Overview of hypoxia around the world. Journal of Environmental Quality 30(2): 275–281.

Dierberg, F.E. and Kiattisimkul, W. (1996). Issues, impacts and implications of shrimp aquaculture in Thailand. Environmental Management 20(5): 649–666.

Dinglasan, M.J.A., Ye, Y., Edwards, E.A. and Mabury, S.A. (2004). Fluorotelomer alcohol biodegradation yields poly- and perfluorinated acids. Environmental Science and Technology 38(10): 2857–2864.

Diver, S. (2006). Aquaponics – integration of hydroponics with aquaculture. ATTRA – National Sustainable Agriculture Information Service, Fayetteville, AR. 28 pp.

Dodds, W.K. (2006). Nutrients and the "dead zone": the link between nutrient ratios and dissolved oxygen in the northern Gulf of Mexico. Frontiers in Ecology and the Environment 4(4): 211–217.

Doney, S.C. (2006). The dangers of ocean acidification. Scientific American 294(3): 58–65.

Donner, S.D., Skirving, W.J., Little, C.M., Oppenheimer, M. and Hoegh-Guldberg, O. (2005). Global assessment of coral bleaching and required rates of adaptation under climate change. Global Change Biology 11: 2251–2265.

Dorenbosch, M. Grol, M.G.G., Nagelkerken, I. and van der Velde, G. (2006). Seagrass beds and mangroves as potential nurseries for the threatened Indo-Pacific humpback wrasse, *Cheilinus undulatus* and Caribbean rainbow parrotfish, *Scarus guacamai*. Biological Conservation 129(2): 277–282.

Dorey, C.N. (2005). A Recipe for Disaster: Supermarkets' Insatiable Appetite for Seafood. Greenpeace Environmental Trust, London.

Douglas, A.E. (2003). Coral bleaching – how and why? Marine Pollution Bulletin 46(4): 385–392.

Duarte, C.M., Marbá, N. and Holmer, M. (2007). Rapid domestication of marine species. Science 316(5823): 382–383.

Earth Negotiations Bulletin (2006). Volume 15 (148): 1–14. International Institute for Sustainable Development (IISD), 13 November 2006.

EC (2003a) European commission technical guidance document on risk assessment, Part II, European Communities, EUR 20418 EN/2. Ch. 3, pp. 93–133. Accessed 2008 at: http://ecb. jrc.it/documents/TECHNICAL_GUIDANCE_DOCUMENT/EDITION_2/tgdpart2_2ed.pdf

EC (2003b) Directive 2003/11/EC of the European Parliament and of the Council of 6 February 2003. Amending for the 24th time Council Directive 76/769/EEC relating to restrictions on the marketing and use of certain dangerous substances and preparations (pentabromodiphenyl ether, octabromodiphenyl ether). Official Journal of the European Union L42, 15.2.2003: 45–46. Accessed 2008 at: http://eurlex.europa.eu/LexUriServ/site/en/oj/2003/l_042/l_04220030215en00450046.pdf

EC (2003c). Directive 2002/95/EC of the European Parliament and of the Council of 27 January 2003 on the restriction of the use of certain hazardous substances in electrical and electronic equipment. Official Journal of the European Union L37, 13.2.2003: 19–23. Accesed 2008 at: http://www.raditek.com/Press%20Releases/Directive2002_95_EC_Jan%202003.pdf

EC (2004). Council Regulation (EC) No 812/2004 of 26.4.2004, laying down measures concerning incidental catches of cetaceans in fisheries and amending Regulation (EC) No 88/98. Official Journal of the European Union L 150, 30.4.2004: 12–31. [http://eurlex.europa.eu/LexUriServ/LexUriServ.do?uri=OJ:L:2004:150:0012:0031:EN:PDF].

EC (2005). Commission Regulation (EC) No 1147/2005 of 15 July 2005, prohibiting fishing for sandeel with certain fishing gears in the North Sea and the Skagerrak. Official Journal of the European Union L185, 16.7.2005: 19. [http://faolex.fao.org/docs/pdf/eur52939.pdf].

EC (2006). Regulation (EC) No 1907/2006 of the European Parliament and of the Council of 18 December 2006 concerning the Registration, Evaluation, Authorisation and Restriction of Chemicals (REACH), establishing a European Chemicals Agency, amending Directive 1999/45/EC and repealing Council Regulation (EEC) No 793/93 and Commission Regulation (EC) No 1488/94 as well as Council Directive 76/769/EEC and Commission Directives 91/155/EEC, 93/67/EEC, 93/105/EC and 2000/21/EC, Official Journal of the European Union L396, 30. 12 2006: 1–848. Accessed 2007 at: http://eurlex.europa.eu/LexUriServ/site/en/oj/2006/l_396/l_39620061230en00010849.pdf

Edinger, E., Jompa, J., Limmon, G.V., Widjatmoko, W. and Risk, M.J. (1998). Reef degradation and coral biodiversity in Indonesia: effects of land-based pollution, destructive fishing practices and changes over time. Marine Pollution Bulletin 36(8): 617–630.

EIA (2006). Stop the Dall's Disaster! Environmental Investigation Agency. London. 4 pp.

Ellis, D.A., Martin, J.W. and Mabury, S.A. (2003). Atmospheric lifetime of fluorotelomer alcohols. Environmental Science and Technology 37(17): 3816–3820.

Ellis, D.A., Martin, J.W., De Silva, A.O., Mabury, S.A., Hurley, M.D., Sulbaek Andersen, M.P. and Wallington, T.J. (2004). Degradation of fluorotelomer alcohols: a likely atmospheric source of perfluorinated carboxylic acids. Environmental Science and Technology 38(12): 3316–3321.

ENDS (2001). Fluorochemicals: 21st century pollutants. Environmental Data Services, ENDS Report 315. p. 23.

ENDS (2003). DuPont in the firing line over fluorochemicals. Environmental Data Services, ENDS Report 340, pp. 9–10.

ENDS (2004). Perfluorinated chemicals: jumping from the frying pan to fire? Environmental Data Services, ENDS Report 354, pp. 28–31.

ENDS (2006). EPA seeks phase-out of PFOA emissions. Environmental Data Services, ENDS Report 373, February.

ENDS (2007). Risk assessments for flame retardants continue. Environmental Data Services, ENDS Report 386, pp. 25–26.

ENDS Europe Daily (2006). Swedes follow through on deca fire-retardant ban. ENDS Europe Daily 2153, 28 August 2006.

ENDS Europe Daily (2008). Norwegians virtually extinguish deca-BDE. ENDS Europe Daily 2465, 18 January 2008.

ENN (2006). 52 new species found on Indonesian reefs. Environmental News Network. Accessed February 2008 at: http://www.coralreef.org/index.php?option=com_content&task=view&id=378&Itemid=14

Environmental Justice Foundation (2003). Smash & Grab: Conflict, Corruption and Human Rights Abuses in the Shrimp Farming Industry. Environmental Justice Foundation, London. 33 pp.

Environmental Justice Foundation (2004). Farming the Sea, Costing the Earth: Why We Must Green the Blue Revolution. Environmental Justice Foundation, London. 77 pp.

Environmental Justice Foundation (2005a). Pirates and Profiteers: How Pirate Fishing Fleets are Robbing People and Oceans. Environmental Justice Foundation, London. 22 pp.

Environmental Justice Foundation (2005b). Party to the Plunder – Illegal Fishing in Guinea and Its Links to the EU. Environmental Justice Foundation, London. 26 pp.

EU (2006) 2,2',6,6'-Tetrabromo-4,4'-isopropylidene diphenol (tetrabromobisphenol-A or TBBP-A). Part II – human health. European Union Risk Assessment Report. European Communities. 170 pp. Accessed 2008 at: http://ecb.jrc.it/DOCUMENTS/Existing-Chemicals/RISK_ASSESSMENT/REPORT/tbbpaHHreport402.pdf

Evans, P.G.H. (1987). The Natural History of Whales & Dolphins. Christopher Helm, London. 343 pp.

FAO (1995). Code of Conduct for Responsible Fisheries. Food and Agricultural Organization of the United Nations, Rome.

FAO (1999). Report of the FAO technical working group meeting on the reduction of incidental catch of seabirds in longline fisheries. Tokyo, Japan, 25–27 March 1998. Food and Agricultural Organization of the United Nations, Rome.

FAO (2003). Status and trends in mangrove area extent worldwide. By Wilkie, M.L. and Fortuna, S. Forest Resources Assessment Working Paper No. 63. Forest Resources Division. FAO, Rome (unpublished). Accessed 2008 at: http://www.fao.org/docrep/007/j1533e/j1533e00.htm

FAO (2004a). State of World Fisheries and Aquaculture 2004. Part 1, World Review of Fisheries and Aquaculture. Food and Agricultural Organization of the United Nations, Rome.

FAO (2004b). International Plan of Action for Reducing Incidental Catch of Seabirds in Longline Fisheries. Food and Agricultural Organization of the United Nations, Rome, 1999.

FAO (2005a). Review of the state of the world marine fishery resources. FAO Fisheries Technical Paper 457. Food and Agricultural Organization of the United Nations, Rome. 20 pp.

FAO (2005b). Report of the technical consultation on sea turtles conservation and fisheries. Bangkok, Thailand, 29 November–2 December 2004. FAO Fisheries Report No. 765. Food and Agricultural Organization of the United Nations, Rome. 32 pp.

FAO (2007a). The State of World Fisheries and Aquaculture 2006. FAO Fisheries and Aquaculture Department. Food and Agricultural Organization of the United Nations, Rome. 162 pp.

FAO (2007b). Report on follow-up actions on sea turtles. Committee on Fisheries Twenty-Seventh Session. Food and Agricultural Organization of the United Nations. Rome, 5–9 March 2007. 8 pp.

FAO (2007c). Cage aquaculture: regional reviews and global overview. FAO Fisheries Technical Paper 498 (eds. M. Halwart, D. Soto and R. Arthur). Food and Agricultural Organization of the United Nations, Rome. 242 pp.

Ferber, D. (2004). Dead zone fix not a dead issue. Science 305: 1557.

Ferguson, S.H., Stirling, I. and McLoughlin, P. (2005). Climate change and ringed seal (*Phoca hispida*) recruitment in Western Hudson Bay. Marine Mammal Science 21(1): 121–135.

Field, C.D. (2000). Mangroves. In: Seas at the Millennium: An Environmental Evaluation. Volume III, Global Issues and Processes (ed. C. Sheppard), Ch. 108, pp. 17–30. Pergamon/Elsevier Science, Oxford. ISBN: 0-08-043207-7.

Fish, M.R., Côté, I.M., Gill, J.A., Jones, A.P., Renshoff, S. and Watkinson, A.R. (2005). Predicting the impact of sea-level rise on Caribbean sea turtle nesting habitat. Conservation Biology 19(2): 482–491.

Fish Site News Desk (2007). Sites close and surveillance increases in Chile's ISA-ravaged waters. The Fish Site News Desk 23 November 2007. Accessed January 2008 at: http://www.thefishsite.com/fishnews/5755/sites-close-and-surveillance-increases-in-chilesisaravaged-waters

Fisheries and Oceans Canada (2003). A scientific review of the potential environmental effects of aquaculture in aquatic ecosystems. Volume 1. Far-field environmental effects of marine finfish aquaculture (B.T. Hargrave). Canadian Technical Report of Fisheries and Aquatic Sciences 2450: ix + 131 pp.

Fisheries and Oceans Canada (2006). Endeavour hydrothermal vents marine protected area. Accessed 2007 at: http://www.pac.dfo-mpo.gc.ca/oceans/mpa/Endeavour_e.htm

Flaherty, M. and Karnjanakesorn, C. (1995). Marine shrimp aquaculture and natural resource degradation in Thailand. Environmental Management 19(1): 27–37.

Fock, F., Uiblein, F., Köster, F. and von Westernhagen, H. (2002). Biodiversity and species-environment relationships of the demersal fish assemblage at the Great Meteor Seamount (subtropical NE Atlantic) sampled by different trawls. Marine Biology 141: 185–199.

Forcada, J., Trathan, P.N., Reid, K., Murphy, E.J. and Croxall, J.P. (2006). Contrasting population changes in sympatric penguin species in association with climate warming. Global Change Biology 12: 411–423.

Fraser, W.R. and Hofmann, E.E. (2003). A predator's perspective on causal links between climate change, physical forcing and ecosystem response. Marine Ecology Progress Series 265: 1–15.

Frederiksen, M., Wanless, S., Harris, M.P., Rothery, P. and Wilson, L.J. (2004). The role of industrial fisheries and oceanographic change in the decline of the North Sea black-legged kittiwakes. Journal of Applied Ecology 41: 1129–1139.

Frei, M. and Becker, K. (2005) Integrated rice-fish culture: coupled production saves resources. Natural Resources Forum 29: 135–143

Fromentin, J-M. and Powers, J.P. (2005). Atlantic bluefin tuna: population dynamics, ecology, fisheries and management. Fish and Fisheries 6: 281–306.

Gabriel, U.U., Akinrotimi, O.A., Anyanwu, P.E., Bekibele, D.O. and Onunkwo, D.N. (2007). The role of dietary phytase in formulation of least cost and less polluting fish feed for sustainable aquaculture development in Nigeria. African Journal of Agricultural Research 2(7): 279–286.

Gage, J.D. and Tyler, P.A. (2001). Deep Sea Biology: A Natural History of Organisms at the Deep-Sea Floor. Cambridge University Press, Cambridge. 504 pp. ISBN: 0-521-33431-4.

Galal, N., Ormond, R.F.G. and Hassan, O. (2002). Effect of a network of no-take reserves in increasing catch per unit effort and stocks of exploited reef fish at Nabq, South Sinai, Egypt. Marine and Freshwater Research 53(2): 199–205.

Galbraith, H., Jones, R., Park, R., Clough, J., Herrod-Julius, S., Harrington, B. and Page, G. (2002). Global climate change and sea level rise: potential losses of intertidal habitat for shorebirds. Waterbirds 25(2): 173–183.

Gales, N.J., Kasuya, T., Clapham, P.J. and Brownell, R.L. (2005). Japan's whaling plan under scrutiny. Nature 435: 883–884.

García Pérez, J.D. (2003). Early socio-political and environmental consequences of the Prestige oil spill in Galicia. Disasters 27(3): 207–223.

Garza-Gil, M.D., Prada-Blanco, A. and Rodríguez, X.V. (2006). Estimating the short-term economic damages from the Prestige oil spill in the Galician fisheries and tourism. Ecological Economics 58: 842–849.

Gaston, A.J., Hipfner, J.M. and Campbell, D. (2002). Heat and mosquitoes cause breeding failures and adult mortality in an Arctic-nesting seabird. IBIS (The International Journal of Avian Science) 144(2): 185–191.

Gatlin, D.M., Barrows, F.T., Brown, P., Dabrowski, K., Gaylord, T.G., Hardy, R.W., Herman, E., Hu, G., Krogdahl, A., Nelson, R., Overturf, K., Rust, M., Sealey, W., Skonberg, D., Souza, E.J., Stone, D., Wilson, R. and Wurtele, E. (2007). Expanding the utilization of sustainable plant products in aquafeeds: a review. Aquaculture Research 38: 551–579.

Gell, F.R. and Roberts, C.M. 2003. Benefits beyond boundaries: the fisheries effects of marine reserves. Trends in Ecology and Evolution 18(9): 448–455.

Gibson, M.A. and Schullinger, S.B. (1998). Answers from the ice edge: the consequences of climate change on life in the Bering and Chukchi seas. Arctic Network and Greenpeace USA (Washington, DC). 32 pp. [http://www.greenpeace.org/international/press/reports/testimonies98].

Giesy, J.P. and Kannan, K. (2002). Perfluorochemical surfactants in the environment. Environmental Science and Technology 36(7): 147A–152A.

Gilchrist, H.G. and Mallory, M.L. (2005). Declines in abundance and distribution of the ivory gull (*Pagophila eburnean*) in Arctic Canada. Biological Conservation 121: 303–309.

Gilman, E., Brothers, N. and Kobayashi, D.R. (2005). Principles and approaches to abate seabird by-catch in longline fisheries. Fish and Fisheries 6: 35–49.

Gilman, E., Moth-Poulsen, T. and Bianchi, G. (2007). Review of measures taken by intergovern-mental organizations to address sea turtle and seabird interactions in marine capture fisheries. FAO Fisheries Circular No. 1025. Food and Agricultural Organization of the United Nations, Rome. 51 pp.

Glowka, L. (2003). Putting marine scientific research on a sustainable footing at hydrothermal vents. Marine Policy 27: 303–312.

Goldburg, R. and Naylor, R. (2005). Future seascapes, fishing, and fish farming. Frontiers in Ecology and the Environment 3(1): 21–28.

Goldburg, R.J., Elliot, M.S. and Naylor, R.L. (2001). Marine aquaculture in the United States. Environmental impacts and policy options. Pew Oceans Commission, Philadelphia, PA. 44 pp.

Goñi, R. (2000). Fisheries effects on ecosystems. In: Seas at the Millennium: An Environmental Evaluation (ed. C.R.C. Sheppard). Volume III, Global Issues and Processes. Ch. 115, pp. 117–133. Elsevier Science, Oxford. ISBN: 0-08-043207-7.

Gräslund, S. and Bengtsson, B.-E. (2001). Chemicals and biological products used in south-east Asian shrimp farming, and their potential impact on the environment – a review. The Science of the Total Environment 280: 93–131.

Grassle, J.F. (1991). Deep-sea benthic biodiversity. Bioscience 41(7): 464–469.

Gray, J.S. (1997). Marine biodiversity: patterns, threats and conservation needs. Biodiversity and Conservation 6: 153–175.

Great Barrier Reef Marine Park Authority (2007). Zoning in the Great Barrier Reef Marine Park. Accessed 2007 at: http://www.gbrmpa.gov.au/__data/assets/pdf_file/0006/19950/ce1207_zoning_q_a.pdf

Green, E.P. and Short, F.T. (2003). World Atlas of Seagrasses. Prepared by the UNEP World Conservation Monitoring Centre. University of California Press, Berkeley, CA. 298 pp.

Greenpeace (1997). From fish to fodder. Accessed 2007 at: http://archive.greenpeace.org/comms/cbio/fodder.html

Greenpeace (1999). Nuclear re-action. Accessed 2007 at: http://www.greenpeace.org.uk/nuclear/nuclear-re-action

Greenpeace (2000). Thousands of radioactive barrels rusting away on the seabed. Press release, 19 June 2000. Accessed 2007 at: http://archive.greenpeace.org/pressreleases/nucreprocess/2000jun19.html

Greenpeace (2001). Witnessing the Plunder: A Report of the MV Greenpeace Expedition to Investigate Pirate Fishing in West Africa. Greenpeace International, Amsterdam, The Netherlands.

Greenpeace (2004a). Polar bears dream of a white Christmas. 28 November 2004. Accessed 2007 at: http://www.greenpeace.org/international/news/polar-bears-dream-of-a-white-c

Greenpeace (2004b). Rescuing the North and Baltic Seas: Marine Reserves - A Key Tool. Greenpeace. 32 pp. ISBN: 1 903907 09 8.

Greenpeace (2005a). Case Studies on IUU Vessels No. 1. The Secret Shame of the *Anuva*. Greenpeace International, Amsterdam, The Netherlands. 8 pp.

Greenpeace (2005b). Case Studies on IUU Vessels No. 2. Chang Xing. Greenpeace International, Amsterdam, The Netherlands. 10 pp.

Greenpeace (2005c). Black Holes in Deep Ocean Space: Closing the Legal Voids in High Seas Biodiversity Protection. Greenpeace International, Amsterdam, The Netherlands. 8 pp.

Greenpeace (2005d). Freedom for the Seas for Now and for the Future: Greenpeace Proposals to Revolutionise High Seas Oceans Governance. Greenpeace International, Amsterdam, The Netherlands. 4 pp.

Greenpeace (2005e). A Recipe for Disaster: Supermarkets' Insatiable Appetite for Seafood. Greenpeace, London. 94 pp.

Greenpeace (2005f). Bioprospecting in the Deep Sea. Greenpeace International, Amsterdam, The Netherlands. 4 pp.

Greenpeace (2006a). Defending the olive ridleys: MV Sugayatri. 16 January 2006. Accessed 2007 at: http://www.greenpeace.org/india/news/turns-turtle

Greenpeace (2006b). Reprocessing. Accessed 2006 at: http://www.greenpeace.org/international/campaigns/nuclear/waste/reprocessing

Greenpeace (2006c). Oil spills – Philippines, Indian Ocean and Lebanon. 18 August 2006. Accessed 2007 at: http://www.greenpeace.org/international/news/recent-oil-spills

Greenpeace (2006d). Comments to "Proposal for a Directive of the European Parliament and of the Council relating to restrictions on the marketing and use of perfluorooctane sulfonates (amendment of the Council Directive 76/769/EEC)", COM (2006) 618 final, 5.12.2005. Prepared by Greenpeace International, January 2006.

Greenpeace (2006e). The binar 4 – a case study of the pirate fishing links between West Africa and Europe. 10 April 2006. 3 pp. Accessed 2007 at: http://www.greenpeace.org/international/press/reports/the-binar-4-a-case-study-of

Greenpeace (2006f). Spain commits to action on stolen fish from West Africa, destined for European plates. 18 April 2006. Accessed 2006 at: http://www.greenpeace.org/international/press/releases/spain-commits-to-action-on-sto

Greenpeace (2006g). Plundering the Pacific. Summary of findings of Greenpeace joint enforcement exercises with FSM and Kiribati, September 4–October 23 2006. Greenpeace International, Amsterdam, The Netherlands. 6 pp.

Greenpeace (2006h). Murky Waters: Hauling in the Net on Europe's High Seas Bottom Trawling Fleet. Greenpeace International, Amsterdam, The Netherlands. 32 pp.

Greenpeace (2006i). Greenpeace Case Study on IUU Fishing # 3. Caught Red-Handed: Daylight Robbery on the High Seas. Greenpeace International, Amsterdam, The Netherlands. 12 pp.

Greenpeace (2006j). Where Have All the Tuna Gone? How Tuna Ranching and Pirate Fishing Are Wiping Out Bluefin Tuna in the Mediterranean Sea. Greenpeace International, Amsterdam, The Netherlands. 40 pp.

Greenpeace (2006k). A Recipe for Change. Supermarkets Respond to the Challenge of Sourcing Sustainable Seafood. Greenpeace, London. 56 pp.

Greenpeace (2006l). The Cod Fishery in the Baltic Sea: Unsustainable and Illegal. Greenpeace International, Amsterdam, The Netherlands. 16 pp.

Greenpeace (2006m). Marine Reserves for the Mediterranean Sea. Greenpeace International, Amsterdam, The Netherlands. 60 pp.

Greenpeace (2006n). Mexico protects the Espiritu Santo Archipelago. 26 November 2006. Accessed 2007 at: http://www.greenpeace.org/international/press/releases/mexico-protects-the-espiritu-s

Greenpeace (2007a). Victory! Sweden (mostly) free from Baltic cod. 7 March 2007. Accessed 2007 at: http://www.greenpeace.org/international/news/victory-sweden-mostly-free

Greenpeace (2007b). The heat is on: the role of marine reserves in boosting ecosystem resilience to climate change. Greenpeace International, Amsterdam, The Netherlands and Greenpeace European Unit, Brussels, Begium. 12 pp. Accessed 2007 at: http://www.greenpeace.org/raw/content/international/press/reports/the-heat-is-on.pdf

Greenpeace (2007c). Regulatory systems for GE crops a failure: the case of MON863. Greenpeace, March 2007. 3 pp. Accessed January 2008 at: http://www.greenpeace.org/raw/content/international/press/reports/gp_briefing_seralini_study.pdf

Greenpeace and Gene Watch UK (2007). GM contamination register. Accessed January 2008 at: http://www.gmcontaminationregister.org

Groenwold, S. and Fonds, M. (2000). Effects on benthic scavengers of discards and damaged benthos produced by the beam-trawl fishery in the southern North Sea. ICES Journal of Marine Science 57: 1395–1406.

G/TBT Notification (2005). PFOS and substances that could degrade to PFOS (perfluroocytl sulphonic derivative), G/TBT Notification Number: G/TBT/N/SWE/51, 06/07/2005: 11 pp. [in Swedish].

Gual, A. (1999). The Bluefin Tuna in the Eastern Atlantic and Mediterranean. Chronicle of a Death Foretold. Greenpeace International, Amsterdam, The Netherlands.

Guard, M. and Masaiganah, M. (1997). Dynamite fishing in Southern Tanzania, geographical variation, intensity of use and possible solutions. Marine Pollution Bulletin 34(10): 758–762.

Haedrich, R.L. and Barnes, S.M. (1997). Changes over time of the size structure in an exploited shelf fish community. Fisheries Research 31: 229–239.

Halim, A. (2002). Adoption of cyanide fishing practice in Indonesia. Ocean and Coastal Management 45: 313–323.

Hall, A.J., Kalantzi, O.I. and Thomas, G.O. (2003). Polybrominated diphenyl ethers (PBDEs) in grey seals during their first year of life – are they thyroid hormone endocrine disrupters? Environmental Pollution 126: 29–37.

Hall, M.A., Alverson, D.L. and Metuzals, K.I. (2000). By-catch: problems and solutions. Marine Pollution Bulletin 41(1–6): 204–219.

Hall, S.J. (1999). By-catch and discards. In: The Effects of Fishing on Marine Ecosystems and Communities. Ch. 2, pp. 16–47. Blackwell Science, Oxford. ISBN: 0-632-04112-9.

Hall, S.J. and Mainprize, B.M. (2005). Managing by-catch and discards: how much progress are we making and how can we do better? Fish and Fisheries 6: 134–155.

Halpern, B.S. (2003). The impact of marine reserves: do reserves work and does reserve size matter? Ecological Applications 13(1): 117–137.

Halpern, B.S., Walbridge, S., Selkoe, K.A., Kappel, C.V., Micheli, F., D'Agrosa, C., Bruno, J.F., Casey, K.S., Ebert, C., Fox, H.E., Fujita, R., Heinemann, D., Lenihan, H.S., Madin, E.M.P., Perry, M.T., Selig, E.R., Spalding, M., Steneck, R. and Watson, R. (2008). A global map of human impact on marine ecosystems. Science 319: 948–952.

Hardy, R. (2007). Farming fish no longer relies only on fish meal feeds. Accessed November 2007 at: http://www.today.uidaho.edu/details.aspx?id=3743

Harley, C.D.G., Hughes, A.R., Hultgren, K.M., Miner, B.G., Sorte, C.J.B., Thornber, C.S., Rodriguez, L.F., Tomanek, L. and Williams, S.L. (2006). The impacts of climate change in coastal marine systems. Ecology Letters 9: 228–241.

Hasan, M.H. (2005). Destruction of a *Holothuria scabra* population by overfishing at Abu Rhamada Island in the Red Sea. Marine Environmental Research 60: 489–511.

Hawkes, L.A., Broderick, A.C., Coyne, M.S., Godfrey, M.H., Lopez-Jurado, L.-F., Lopez-Suarez, P., Merino, S.E., Varo-Cruz, N. and Godley, B.J. (2006). Phenotypically linked dichotomy in sea turtle foraging requires multiple conservation approaches. Current Biology 16: 990–995.

Hays, G.C., Richardson, A.J. and Robinson, C. (2005). Climate change and marine plankton. Trends in Ecology and Evolution 20(6): 337–344.

Heilman, K. (2006). Nautilus one step closer to undersea mining. Accessed 2007 at: http://www.resourceinvestor.com/pebble.asp?relid=24459

Hekster, F.M., Laane, R.W.P.M. and de Voogt, P. (2003). Environmental and toxicity effects of perfluoroalkylated substances. Reviews of Environmental Contamination and Toxicology 179: 99–121.

Hemphill, A.H. (2005). Conservation on the high seas – drift algae habitat as an open ocean cornerstone. Parks Magazine of the IUCN (World Conservation Union) 15(3): 48–56.

High Seas Task Force (2006). Closing the net: stop illegal fishing on the high seas. Governments of Australia, Canada, Chile, Namibia, New Zealand, and the United Kingdom, WWF, IUCN and the Earth Institute at Columbia University.

Hilborn, R., Branch, T.A., Ernst, B., Magnusson, A., Minte-Vera, C.V., Scheuerell, M.D. and Valero, J.L. (2003). State of the world's fisheries. Annual Review of Environment and Resources 28: 359–399.

Hindar, K. and Diserud, O. (2007). Sårbarhetsvurdering av ville laksebestander overfor rømtopp-drettslaks. NINA Rapport 244 (Norsk Institutt for Naturforskning). 50 pp.

Hobday, A.J., Tegner, M.J. and Haaker, P.L. (2001). Over-exploitation of a broadcast spawning marine invertebrate: decline of the white abalone. Reviews in Fish Biology and Fisheries 10: 493–514.

Hodgson, G. (1999). A global assessment of human effects on coral reefs. Marine Pollution Bulletin 38(5): 345–355.

Hoegh-Guldberg, O. (1999). Climate change, coral bleaching and the future of the world's coral reefs. Marine and Freshwater Ecology 50: 839–866.

Hoegh-Guldberg, O. (2005). Low coral cover in a high-CO_2 world. Journal of Geophysical Research 110 (24 August): Art No. C09S06.

Holmström K., Gräslund, S., Wahlström, A., Poungshompoo, S., Bengtsson, B.-E. and Kautsky, N. (2003). Antibiotic use in shrimp farming and implications for environmental impacts and human health. International Journal of Food Science and Technology 38: 255–266.

Holmström, K.E., Järnberg, U. and Bignert, A. (2005). Temporal trends of PFOS and PFOA in guillemot eggs from the Baltic Sea, 1968–2003. Environmental Science and Technology 39(1): 80–84.

Holt, S. (2006). Propaganda and pretext. Marine Pollution Bulletin 52: 363–366.

Holt, S.J. (2002). The whaling controversy. Fisheries Research 54: 145–151.

Hooper, J., Clark, J.M., Charman, C. and Agnew, D. (2005). Seal mitigation measures on trawl vessels fishing for krill in CCAMLR subarea 48.3. CCAMLR Science 12: 195–205.

Houde, M., Bujas, T.A.D., Small, J., Wells, R.S., Fair, P.A., Bossart, G.D., Solomon, K.R. and Muir, D.C.G. (2006a). Biomagnification of perfluoroalkyl compounds in bottlenose dolphin (Tursiops truncates) food web. Environmental Science and Technology 40: 4138–4144.

Houde, M., Martin, J.W., Letcher, R.J., Solomon, K.R. and Muir, D.C.G (2006b). Biological monitoring of polyfluoralkyl substances. Environmental Science and Technology 40(11): 3463–3473.

Hoyt, E. (2001). Whale Watching 2001: Worldwide Tourism Numbers, Expenditures and Expanding Socioeconomic Benefits. International Fund for Animal Welfare, Yarmouth Port, MA. ISBN: 1-901002-09-8.

Hu, W., Jones, P.D., Upham, B.L., Trosko, J.E., Lau, C. and Giesy, J.P. (2002). Inhibition of gap junctional intercellular communication by perfluorinated compounds in rat liver and dolphin kidney cells in vitro and Sprague-Dawley rats in vivo. Toxicological Sciences 68(2): 429–436.

Hughes, L. (2003). Climate change and Australia: trends, projections and impacts. Austral Ecology 28: 423–443.

Hughes, R.G. (2004). Climate change and loss of saltmarshes: consequences for birds. IBIS (The International Journal of Avian Science) 146(Suppl. 1): 21–28.

Hughes, T.P., Bellwood, D.R. and Connolly, S.R. (2002). Biodiversity hotspots, centres of endemicity, and the conservation of coral reefs. Ecology Letters 5: 775–784.

Hughes, T.P., Baird, A.H., Bellwood, D.R., Card, M., Connolly, S.R., Folke, C., Grosberg, R., Hoegh-Guldberg, O., Jackson, J.B.C., Kleypas, J., Lough, J.M., Marshall, P., Nyström, M., Palumbi, S.R., Pandolfi, J.M., Rosen, B. and Roughgarden, J. (2003). Climate change, human impacts, and the resilience of coral reefs. Science 301: 929–933.

Huntington, H. and Fox, S. (2004). The changing Arctic: indigenous perspectives. In: ACIA (Arctic Climate Impact Assessment) Scientific Report. Ch. 3, pp. 61–98. Cambridge University Press, Cambridge.

Huntington, T.C. (2004a). Feeding the fish: sustainable fish feed and Scottish aquaculture. Report to the Joint Marine Programme (Scottish Wildlife Trust and WWF Scotland) and RSPB Scotland. Poseiden Aquatic Resource Management, Hampshire, UK. 49 pp.

Huntington, T.C. (2004b). Assessment of the sustainability of industrial fisheries producing fish meal and fish oil. Final report to the Royal Society for the Protection of Birds by Poseidon Aquatic Resource Management Ltd and the University of Newcastle upon Tyne. Poseidon Aquatic Resource Management, Hampshire, UK. 105 pp.

Hutchings, J.A. (2000). Collapse and recovery of marine fishes. Nature 406: 882–885.

De la Huz, R., Lastra, M., Junoy, J., Castellanos, C. and Viéitez, J.M. (2005). Biological impacts of oil pollution and cleaning in the intertidal zone of exposed sandy beaches: preliminary study of "Prestige" oil spill. Estuarine, Coastal and Shelf Science 65: 19–29.

Hyrenbach, K.D. and Dotson, R.C. (2003). Assessing the susceptibility of female black-footed albatross (Phoebastria nigripes) to longline fisheries during their post-breeding dispersal: an integrated approach. Biological Conservation 112: 391–404.

IAC (2006). Inter-American Convention for the protection and conservation of sea turtles. Accessed 2007 at: http://www.iacseaturtle.org/iacseaturtle/English/home.asp

ICES (2006). Overhaul Deepsea Fisheries, Sharks in Trouble, Good and Bad News for Other Fish Stocks. Press Release, International Council for the Exploration of the Sea, 17 October 2005.

ICES (2007). Report of the ICES Advisory Committee on Fishery Management, Advisory Committee on the Marine Environment and Advisory Committee on Ecosystems, 2007. ICES Advice. Book 6, 249 pp.

International Technical Expert Workshop on Marine Turtle By-catch in Longline Fisheries (2003). Seattle, Washington, DC, 11–13 February 2003.

IPCC (2001a). Climate Change 2001: The Scientific Basis. Contribution of Working Group I to the Third Assessment Report of the Intergovernmental Panel on Climate Change (eds. J.T. Houghton, Y. Ding, D.J. Griggs, M. Noguer, P.J. van der Linden, X. Dai, K. Maskell and C.A. Johnson), 881 pp. Cambridge University Press, Cambridge, UK/New York.

IPCC (2001b). Coastal zones and marine ecosystems, Chapter 6. In: Climate Change 2001: Impacts, Adaptation and Vulnerability. Contribution of Working Group II to the Third Assessment Report of the Intergovernmental Panel on Climate Change (eds. J.J. McCarthy, O.F. Canziani, N.A. Leary, D.J. Dokken and White K.S.), Ch. 6, pp. 345–379. Cambridge University Press, Cambridge, UK/New York.

IPCC (2007a). Summary for policy makers. In: Climate Change 2007: The Physical Science Basis. Contribution of Working Group I to the Fourth Assessment Report of the Intergovernmental Panel on Climate Change (eds. S. Solomon, D. Qin, M. Manning, Z. Chen, M. Marquis, K.B. Averyt, M. Tignor and H.L. Miller), 18 pp. Cambridge University Press, Cambridge, UK/New York.

IPCC. Solomon, S., Qin, D., Manning, M., Alley, R.B., Berntsen, T., Bindoff, N.L., Chen, Z., Chidthaisong, A., Gregory, J.M., Hegerl, G.C., Heimann, M., Hewitson, B., Hoskins, B.J., Joos, F., Jouzel, J., Kattsov, V., Lohmann, U., Matsuno, T., Molina, M., Nicholls, N., Overpeck, J., Raga, G., Ramaswamy, V., Ren, J., Rusticucci, M., Somerville, R., Stocker, T.F., Whetton, P., Wood, R.A. and Wratt, D. (2007b) Technical summary. In: Climate Change 2007: The Physical Science Basis. Contribution of Working Group I to the Fourth Assessment Report of the Intergovernmental Panel on Climate Change (eds. S. Solomon, D. Qin, M. Manning, Z. Chen, M. Marquis, K.B. Averyt, M. Tignor and H.L. Miller), 73 pp. Cambridge University Press, Cambridge, UK/New York.

IPCC, Bindoff, N.L., Willebrand, J., Artale, V., Cazenave, A., Gregory, J., Gulev, S., Hanawa, K., Le Quéré, C., Levitus, S., Nojiri, Y., Shum, C.K., Talley, L.D. and Unnikrishnan, A. (2007c) Observations: oceanic climate change and sea level. In: Climate Change 2007: The Physical Science Basis. Contribution of Working Group I to the Fourth Assessment Report of the Intergovernmental Panel on Climate Change (eds. S. Solomon, D. Qin, M. Manning, Z. Chen, M. Marquis, K.B. Averyt, M. Tignor and H.L. Miller), Ch. 5, pp. 385–432. Cambridge University Press, Cambridge, UK/New York [http://ipcc-wg1.ucar.edu/wg1/wg1-report.html].

IPCC, Lemke, P., Ren, J., Alley, R.B., Allison, I., Carrasco, J., Flato, G., Fujii, Y., Kaser, G., Mote, P., Thomas, R.H. and Zhang, T. (2007d) Observations: changes in snow, ice and frozen ground. In: Climate Change 2007: The Physical Science Basis. Contribution of Working Group I to the Fourth Assessment Report of the Intergovernmental Panel on Climate Change (eds. S. Solomon, D. Qin, M. Manning, Z. Chen, M. Marquis, K.B. Averyt, M. Tignor and H.L. Miller), Ch. 4, pp. 337–383. Cambridge University Press, Cambridge, UK/New York.

IPCC, Nicholls, R.J., Wong, P.P., Burkett, V.R., Codignotto, J.O., Hay, J.E., McLean, R.F., Ragoonaden, S. and Woodroffe, C.D. (2007e) Coastal systems and low-lying areas. In: Climate Change 2007: Impacts, Adaptation and Vulnerability. Contribution of Working Group II to the Fourth Assessment Report of the Intergovernmental Panel on Climate Change (eds. M.L. Parry, O.F. Canziani, J.P. Palutikof, P.J. van der Linden and C.E. Hanson), Ch. 6, pp. 315–356. Cambridge University Press, Cambridge.

Irvine, G.V., Mann, D.H. and Short, J.W. (2006). Persistence of 10-year old Exxon Valdez oil on Gulf of Alaska beaches. The importance of boulder-armouring. Marine Pollution Bulletin 52(9): 1011–1022.

Islam, M.S. and Haque, M. (2004). The mangrove-based coastal and nearshore fisheries of Bangladesh: ecology, exploitation and management. Reviews in Fish Biology and Fisheries 14: 153–180.

Islam, M.S., Wahad, M.A and Tanaka, M. (2004). Seed supply for coastal brackish water shrimp farming: environmental impacts and sustainability. Marine Pollution Bulletin 48: 7–11.

IUCN (2006). 2006 IUCN red list of threatened species. World Conservation Union. Accessed 2006 at: http://www.redlist.org/

Iwama, G.K. (1991). Interactions between aquaculture and the environment. Critical Reviews in Environmental Control 21(2): 177–216.

Jackson, J.B.C., Kirby, M.X., Berger, W.H., Bjorndal, K.A., Botsford, L.W., Bourque, B.J., Bradbury, R.H., Cooke, R., Ealandson, J., Estes, J.A., Hughes, T.P., Kidwell, S., Lange, C.B., Lenihan, H.S., Pandolfi, J.M., Peterson, C.H., Steneck, R.S., Tegner, M.J. and Warner, R.R. (2001a). Historical overfishing and the recent collapse of coastal ecosystems. Science 293: 629–638.

Jackson, E.L., Rowden, A.A., Attrill, M.J., Bossey, S.J. and Jones, M.B. (2001b). The importance of seagrass beds as a habitat for fishery species. Oceanography and Marine Biology: An Annual Review 39: 269–303.

Janák, K., Covaci, A., Voorspoels, S. and Becher, G. (2005). Hexabromocyclododecane in marine species from the Western Scheldt estuary: diastereoisomer- and enantiomer-specific accumulation. Environmental Science and Technology 39: 1987–1994.

Jennings, S., Pinnegar, J.K., Polunin, N.V.C. and Warr, K.J. (2001). Impacts of trawling disturbance on the trophic structure of benthic invertebrate communities. Marine Ecology Progress Series 213: 127–142.

Johansson, I., Héas-Moisan, K., Guiot, N., Munschy, C. and Tronczyński, J. (2006). Polybrominted diphenyl ethers (PBDEs) in mussels from selected French coastal sites: 1981–2003. Chemosphere 64: 296–305.

Johnson, A., Salvador, G., Kenny, J., Robbins, J., Kraus, S., Landry, S. and Clapham, P. (2005). Fishing gear involved in entanglements of right and humpback whales. Marine Mammal Science 21(4): 635–645.

Johnson-Restrepo, B., Kannan, K., Addink, R. and Adams, D.H. (2005). Polybrominated diphenyl ethers and polychlorinated biphenyls in a marine foodweb of coastal Florida. Environmental Science and Technology 39: 8243–8250.

Johnston, P. and Santillo, D. (2000). Whales in competition with commercial fisheries: a modern myth based on pseudo-science. Greenpeace Research Laboratories, University of Exeter, UK. Technical Note 05/00. 10 pp.

Johnston, P., Santillo, D., Stringer, R., Ashton, J. McKay, B., Verbeek, M., Jackson, E., Landman, J., van den Broek, J., Samsom, D. and Simmonds, M. (1998). Greenpeace report on the world's oceans. Greenpeace Research Laboratories Report, May 1998. Stichting Greenpeace Council, Amsterdam, The Netherlands. 154 pp. ISBN: 90-73361-45-1.

Johnston, P., Santillo, D., Ashton, J. and Stringer, R. (2000). Sustainability of human activities on marine ecosystems. In: Seas at the Millennium: An Environmental Evaluation. Volume III, Global Issues and Processes (eds. C. Sheppard), Ch. 133, pp. 359–374. Pergamon/Elsevier Science, Oxford. ISBN: 0-08-043207-7.

Johnston, P.A. and Santillo, D. (2004). Conservation of seamount ecosystems: application of a marine protected areas concept. Archive of Fishery and Marine Research 51(1–3): 305–319.

Jokiel, P. and Brown, E.K. (2004). Global warming, regional trends and inshore environmental conditions influence coral bleaching in Hawaii. Global Change Biology 10: 1627–1641.

Kaiser, M.J. and Spencer, B.E. (1996). The effects of beam-trawl disturbance on infaunal communities in different habitats. The Journal of Animal Ecology 65(3): 348–358.

Kaiser, M.J., Edwards, D.B., Armstrong, P.J., Radford, K., Lough, N.E.L., Flatt, R.P. and Jones, H.D. (1998). Changes in megafaunal benthic communities in different habitats after trawling disturbance. ICES Journal of Marine Science 55: 353–361.

Kajiwara, N., Ueno, D., Takahashi, A., Baba, N. and Tanabe, S. (2004). Polybrominated diphenyl ethers and organochlorines in archived northern fur seal samples from the Pacific coast of Japan. Environmental Science and Technology 38(14): 3804–3809.

Kalantzi, O.I., Hall, A.J., Thomas, G.O. and Jones, K.C. (2005). Polybrominated diphenyl ethers and selected organochlorine chemicals in grey seals (*Halichoerus grypus*) in the North Sea. Chemosphere 58: 345–354.

Kannan, K., Franson, J.C., Bowerman, W.W., Hansen, K.J., Jones, P.D. and Giesy, J.P. (2001a). Perfluorooctane sulphonate in fish-eating water birds including bald eagles and albatrosses. Environmental Science and Technology 35(15): 3065–3070.

Kannan, K., Koistinen, J., Beckmen, K., Evans, T., Gorzelany, J.F., Hansen, K.J., Jones, P.D., Helle, E., Nyman, M. and Giesy, J.P. (2001b). Accumulation of perfluorooctane sulphonate in marine mammals. Environmental Science and Technology 35(8): 1593–1598.

Kannan, K., Corsolini, S., Falandysz, J., Oehme, G., Focardi, S. and Giesy, J.P. (2002a). Perfluorooctanesulfonate and related fluorinated hydrocarbons in marine mammals, fishes, and birds from coasts of the Baltic and the Mediterranean Seas. Environmental Science and Technology 36(15): 3210–3216.

Kannan, K., Hansen, K.J., Wade, T.L. and Giesy, J.P. (2002b). Perfluorooctane sulphonate in oysters, *Crassostrea virginica*, from the Gulf of Mexico and the Chesapeake Bay, USA. Archives of Environmental Contamination and Toxicology 42: 313–318.

Kannan, K., Choi, J.-W., Iseki, N., Senthilkumar, K., Kim, D.H., Masunaga, S. and Giesy, J.P. (2002c). Concentrations of perfluorinated acids in livers of birds from Japan and Korea. Chemosphere 49: 225–231.

Kannan, K., Perrotta, E. and Thomas, N.J. (2006). Association between perfluorinated compounds and pathological conditions in southern sea otters. Environmental Science and Technology 40(16): 4943–4948.

Karlson, K., Rosenberg, R. and Bonsdorff, E. (2002). Temporal and spatial large-scale effects of eutrophication and oxygen deficiency on benthic fauna in Scandinavian and Baltic waters – a review. Oceanography and Marine Biology: An Annual Review 40: 427–489.

Kaschner, K. and Pauly, D. (2004). Competition Between Marine Mammals and Fisheries: Food for Thought. Published by the Humane Society of the United States/Humane Society International.

Kasuya, T. (2007). Japanese whaling and other cetacean fisheries. Environmental Science and Pollution Research 14(1): 39–48.

Kathiresan, K. and Rajendran, N. (2002). Fishery resources and economic gain in three mangrove areas on the south-east coast of India. Fisheries Management and Ecology 9: 277–283.

Kaunda-Arara, B. and Rose, G.A. (2004). Effects of marine reef national parks on fishery CPUE in coastal Kenya. Biological Conservation 118: 1–13.

Kelly, B.C., Gobas, F.A.C. and McLachlan, M.S. (2004). Intestinal absorption and biomagnification of organic contaminants in fish, wildlife and humans. Environmental Toxicology and Chemistry 23(10): 2324–2336.

Kershaw, P.J., McCubbin, D. and Leonard, K.S. (1999). Continuing contamination of north Atlantic and Arctic waters by Sellafield radionuclides. The Science of the Total Environment 237/238: 119–132.

Key, B.D., Howell, R.D. and Criddle, C.S. (1997). Fluorinated organics in the biosphere. Environmental Science and Technology 31(9): 2445–2454.

Kitchingman, A. and Lai, S. (2004). Inferences on potential seamount locations from mid-resolution bathymetric data. In: Seamounts: Biodiversity and Fisheries (eds. T. Morato and D. Pauly). Fisheries Centre Research Reports 12(5): 7–12. Fisheries Centre, University of British Columbia, Canada.

Kleypas, J.A., Buddemeier, R.W., Archer, D., Gauttuso, J.-P., Langdon, C. and Opdyke, N. (1999). Geochemical consequences of increased atmospheric carbon dioxide on coral reefs. Science 284(5411): 118–120.

Knauss, J.A. (1997) The International Whaling Commission - its past and possible future. Ocean Development and International Law 28: 79–87

Knowlton, N. (2001). The future of coral reefs. Proceedings of the National Academy of Sciences of the United States of America 98(10): 5419–5425.

Koch, V., Nichols, W.J., Peckham, H. and de la Toba, V. (2006). Estimates of sea turtle mortality from poaching and by-catch in Bahía Magdalena, Baja California Sur, Mexico. Biological Conservation 128: 327–334.

Koslow, J.A., Williams, A. and Paxton, J.R. (1997). How many demersal fish species in the deep sea? A test of a method to extrapolate from local to global diversity. Biodiversity and Conservation 6: 1523–1532.

Koslow, J.A., Boehlert, G.W., Gordon, J.D.M., Haedrich, R.L., Lorance, P. and Parin, N. (2000). Continental slope and deep-sea fisheries: implications for a fragile ecosystem. ICES Journal of Marine Science 57: 548–557.

Koslow, J.A., Gowlett-Holmes, K., Lowry, J.K., O'Hara, T., Poore, G.C.B. and Williams, A. (2001). Seamount benthic macrofauna off southern Tasmania: community structure and impacts of trawling. Marine Ecology Progress Series 213: 111–125.

Kraus, S.D., Brown, M.W., Caswell, H., Clark, C.W., Fujiwara, M., Hamilton, P.K., Kenney, R.D., Knowlton, A.R., Landry, S., Mayo, C.A., McLellan, W.A., Moore, M.J., Nowacek, D.P., Pabst, A.A., Read, A.J. and Rolland, R.M. (2005). North Atlantic right whales in crisis. Science 309: 561–562.

Kristofersson, D. and Anderson, J.L. (2006). Is there a relationship between fisheries and farming? Interdependence of fisheries, animal production and aquaculture. Marine Policy 30: 721–725

Krkošek, M., Lewis, M.A., Morton, A., Frazer, L.N. and Volpe, J.P. (2006). Epizootics of wild fish induced by farm fish. PNAS 103(42): 15506–15510.

Krkošek, M., Ford, J.S., Morton, A., Lele, S., Myers, R.A. and Lewis, M.A. (2007). Declining wild salmon populations in relation to parasites from farm salmon. Science 318(5857): 1772–1775.

Langdon, C. and Aitkinson, M.J. (2004). Reduction of coral calcification from CO_3^{2-} decreases by the mid-21st century. Symposium on the oceans in a high-CO_2 world. 10–12 May 2004, UNESCO, Paris.

Lau, C., Thibodeaux, J.R., Hanson, R.G., Rogers, J.M., Grey, B.E., Stanton, M.E., Butenhoff, J.L. and Stevenson, L.A. (2003). Exposure to perfluorooctane sulphonate during pregnancy in rat and mouse. II: postnatal evaluation. Toxicological Sciences 74(2): 382–392.

Lau, C., Butenhoff, J.L. and Rogers, J.M. (2004). The developmental toxicity of perfluoroalkyl acids and their derivatives. Toxicology and Applied Pharmacology 198(2): 231–241.

Law, R.J., Alaee, M., Allchin, C.R., Boon, J.P., Lebeuf, M., Lepom, P. and Stern, G.A. (2003). Levels and trends of polybrominated diphenylethers and other brominated flame retardants in wildlife. Environment International 29: 757–770.

Law, R.J., Allchin, C.R., de Boer, J., Covaci, A., Herzke, D., Lepom, P., Morris, S., Tronczynski, J. and de Wit, C.A. (2006a). Levels and trends of brominated flame retardants in the European environment. Chemosphere 64(2): 187–208.

Law, R.J., Bersuder, P., Allchin, C.R. and Barry, J. (2006b). Levels of the flame retardants hexabromocyclododecane and the tetrabromobisphenol A in the blubber of harbour porpoises (Phocoena phocoena) stranded or bycaught in the U.K., with evidence for an increase in HBCD concentrations in recent years. Environmental Science and Technology 40(7): 2177–2183.

Le, T.X., Munekage, Y. and Shin-ichiro, K. (2005). Antibiotic resistance in bacteria from shrimp farming in mangrove areas. The Science of the Total Environment 349: 95–105.

Lear, W.H. (1998). History of fisheries in the Northwest Atlantic: the 500 year perspective. Journal of Northwest Atlantic Fisheries Science 23: 41–73.

Legler, J. and Brouwer, A. (2003). Are brominated flame retardants endocrine disruptors? Environment International 29: 879–885.

Leung, A.O.W., Luksemburg, W.J., Wong, A.S. and Wong, M.H. (2007). Spatial distribution of polybrominated diphenyl ethers and polychlorinated dibenzo-p-dioxins and dibenzofurans in soil and combusted residue at Guiyu, an electronic waste recycling site in Southeast China. Environmental Science and Technology 41(8): 2730–2737.

Lewison, R.L., Crowder, L.B. and Shaver, D.J. (2003). The impact of turtle excluder devices and fisheries closures on loggerhead and Kemp's ridley strandings in the western Gulf of Mexico. Conservation Biology 17(4): 1089–1097.

Lewison, R.L., Crowder, L.B., Read, A.J. and Freeman, S.A. (2004a). Understanding impacts of fisheries by-catch on marine megafauna. Trends in Ecology and Evolution 19(11): 598–604.

Lewison, R.L., Freeman, S.A. and Crowder, L.B. (2004b). Quantifying the effects of fisheries on threatened species: the impact of pelagic longlines on loggerhead and leatherback sea turtles. Ecology Letters 7: 221–231.

Lindahl, P., Ellmark, C., Gäfvert, T., Mattson, S., Roos, P., Holm, E. and Erlandsson, B. (2003). Long term study of ^{99}Tc in the marine environment on the Swedish west coast. Journal of Environmental Radioactivity 67: 145–156.

Little, C.T.S. and Vrijenhoek, R.C. (2003). Are hydrothermal vent animals living fossils? Trends in Ecology and Evolution 18(11): 582–588.

Loeb, V., Siegel, V., Holm-Hansen, O., Hewitt, R., Fraser, W., Trivelpiece, W. and Trivelpiece, S. (1997). Effects of sea-ice extent and krill or salp dominance on the Antarctic food web. Nature 387: 897–900.

Loeng, H. (2004). Marine systems. In: ACIA (Arctic Climate Impact Assessment) Scientific Report. Ch. 9, pp. 453–538. Cambridge University Press, Cambridge.

Losada, S. (2007). Pirate Booty: How ICCAT Is Failing to Curb IUU Fishing. Greenpeace, Madrid. 50 pp.

Lovatelli, A. (2005). Summary report of the status of bluefin tuna aquaculture in the Mediterranean. FAO Fisheries Report No. 779. Food and Agricultutal Organization of the United Nations, Rome.

Lyle-Fritch, L.P., Romero-Beltrán, E. and Páez-Osuna, F. (2006). A survey on use of the chemical and biological products for shrimp farming in Sinaloa (NW Mexico). Aquacultural Engineering 35: 135–146.

Lynnes, A.S., Reid, K. and Croxall, J.P. (2004). Diet and reproductive success of Adélie and chin-strap penguins: linking response of the predators to prey population dynamics. Polar Biology 27: 544–554.

Machias, A., Vassilopoulou, V., Vatsos, D., Bekas, P., Kallianiotis, A., Papaconstantinou, C. and Tsimenides, N. (2001). Bottom trawl discards in the northeastern Mediterranean. Fisheries Research 53: 181–195.

Malakoff, D. (2004). New tools reveal treasures at ocean hot spots. Science 304(5674): 1104–1105.

Marigómez, I., Soto, M., Cancio, I., Orbea, A., Garmendia, L. and Cajaraville, M.P. (2006). Cell and tissue biomarkers in mussel, and histopathology in hake and anchovy from Bay of Biscay after the Prestige oil spill (Monitoring Campaign 2003). Marine Pollution Bulletin 53(5–7): 287–304.

Marine Conservation Unit (2005). Protecting our seas – Tiakina a Tangaroa. An overview of New Zealand's marine biodiversity conservation and the role of marine protected areas. Published by Marine Conservation Unit, Department of Conservation, Wellington. 13 pp. [http://www.doc.govt.nz/templates/MultiPageDocumentTOC.aspx?id=39741].

Marquez, J.V. (2008). The human rights consequences of inequitable trade and development expansion: abuse of law and community rights in the Gulf of Fonseca, Honduras. Accessed January 2008 at: http://www.mangroveactionproject.org/issues/shrimp-farming/shrimp-farming

Marshall, P., Airame, S., Obura, D. and Maitland, P. (2005). Climate change and ocean warming: preparing MPAs for it. MPA News 6(8): 1–3.

Martin, J.W., Mabury, S.A., Solomon, K.R. and Muir, D.C.G. (2003). Bioconcentration and tissue distribution of perfluorinated acids in rainbow trout. Environmental Toxicology and Chemistry 22(1): 196–204.

Martin, J.W., Smithwick, M.M., Braune, B.M., Hoekstra, P.F., Muir, D.C.G. and Mabury, S.A. (2004a). Identification of long-chain perfluorinated acids in biota from the Canadian Arctic. Environmental Science and Technology 38(2): 373–380.

Martin, M., Lam, P.K.S. and Richardson, B.J. (2004b). An Asian quandary: where have all the PBDEs gone? Marine Pollution Bulletin 49: 375–382.

Martínez-Abraín, A., Velando, A., Oro, D., Genovart, M., Gerique, C., Angel Bartolomé, M., Villuendas, E. and Sarzo, B. (2006). Sex-specific mortality of European shags after the

Prestige oil spill: demographic implications for the recovery of colonies. Marine Ecology Progress Series 318: 271–276.

McClanahan, T.R. and Arthur, R. (2001). The effect of marine reserves and habitat on populations of East African coral reefs. Ecological Applications 11(2): 559–569.

McCubbin, D., Leonard, K.S., McDonald, P., Bonfield, R. and Boust, D. (2006). Distribution of Technetium-99 in subtidal sediments of the Irish Sea. Continental Shelf Research 26: 458–473.

McDiarmid, B., Gotje, M. and Sack, K. (2005). NAFO Case Study. The Northwest Atlantic Fisheries Organisation: A Case Study in How RFMOs Regularly Fail to Manage Our Oceans. Greenpeace International, Amsterdam, The Netherlands. 21 pp.

McGarvin, M. (2005). Deep-Water Fishing: Time to Stop the Destruction. Greenpeace International, Amsterdam, The Netherlands. 19 pp.

McManus, J.W., Meñez, L.A.B., Kesner-Reyes, K.N., Vergara, S.G. and Ablan, M.C. (2000). Coral reef fishing and coral-algal phase shifts: implications for global reef status. ICES Journal of Marine Science 57: 572–578.

McSmith, A. (2006). Poisonous legacy of Buncefield fire. The Independent, 5 May 2006.

Mensink, B.P., Fischer, C.V., Cadée, G.C., Fonds, M., Ten Hallers-Tjabbes, C.C. and Boon, J.P. (2000). Shell damage and mortality in the common whelk *Buccinum undatum* caused by beam trawl fishery. Journal of Sea Research 43: 53–64.

Mente, E., Pierce, G.J., Santos, M.B. and Neofitou, C. (2006). Effect of feed and feeding in the culture of salmonids on the marine aquatic environment: a synthesis for European aquaculture. Aquaculture International 14: 499–522.

Meredith, M.P. and King, J.C. (2005). Rapid climate change in the ocean west of the Antarctic Peninsula during the second half of the 20th century. Geophysical Research Letters 32: L19604.

Miles, R.D. and Chapman, F.A. (2006). The benefits of fish meal in aquaculture diets. Document FA122, one of a series of the Department of Fisheries and Aquatic Sciences. Florida Cooperative Extension Service, Institute of Food and Agricultural Sciences, University of Florida, Florida. First published in May 2006.

Millennium Ecosystem Assessment (2005). Ecosystems and Human Wellbeing: Current State and Trends, Volume 1 (eds. R. Hassan, R. Scholes and N. Ash). Island Press, Washington/ Covelo, London.

Miranda, C.D. and Zemelman, R. (2002). Bacterial resistance to oxytetracycline in Chilean salmon farming. Aquaculture 212: 31–47.

Moberg, F. and Folke, C. (1999). Ecological goods and services of coral reef ecosystems. Ecological Economics 29: 215–233.

Moline, M.A., Claustre, H., Frazer, T.K., Schofields, O. and Vernet, M. (2004). Alteration of the food web along the Antarctic Peninsula in response to a regional warming trend. Global Change Biology 10: 1973–1980.

Moody, C.A. and Field, J.A. (1999). Determination of perfluorocarboxylates in groundwater impacted by fire-fighting activity. Environmental Science and Technology 33(16): 2800–2806.

Mora, C., Andréfouët, S., Costello, M.J., Kranenburg, C., Rollo, A., Veron, J., Gaston, K.J. and Myers, R.A. (2006). Coral reefs and the global network of marine protected areas. Science 312: 1750–1751.

Morales-Caselles, C., Jiménez-Tenorio, N., de Canales, M.L.G., Sarasquete, C. and DelValls, T.A. (2006). Ecotoxcity of sediments contaminated by the oil spill associated with the tanker "Prestige" using juveniles of fish (*Sparus aurata*). Archives of Environmental Contamination Toxicology 51: 652–660.

Morgera, E. (2004). Whale sanctuaries: an evolving concept within the international whaling commission. Ocean Development and International Law 35: 319–338.

Morris, S., Allchin, C.R., Zegers, B.N., Haftka, J.J.H., Boon, J.P., Belpaire, C., Leonards, P.E.G., van Leeuwen, S.P.J. and de Boer, J. (2004). Distribution and fate of HBCD and TBBPA brominated flame retardants in North Sea estuaries and aquatic food webs. Environmental Science and Technology 38: 5497–5504.

MPA news (2005). Fiji designates five MPAs as part of network. MAP News 7(5): 2.

Muir, D.C.G., Backus, S., Derocher, A.E., Dietz, R., Evans, T.J., Gabrielsen, G.W., Nagy, J., Norstrom, R.J., Sonne, C., Stirling, I., Taylor, M.K. and Letcher, R.J. (2006). Brominated flame retardants in polar bears (*Ursus maritimus*) from Alaska, the Canadian Arctic, East Greenland, and Svalbard. Environmental Science and Technology 40: 449–455.

Mumby, P.J., Edwards, A.J., Arias-González, J.E., Lindeman, K.C., Blackwell, P.G., Gall, A., Gorczynska, M.I., Harbourne, A.R., Pescod, C.L., Renken, H., Wabnitz, C.C.C. and Llewellyn, G. (2004). Mangroves enhance the biomass of coral reef fish communities in the Caribbean. Nature 427: 533–536.

Mullon, C., Fréon, P. and Cury, P. (2005). The dynamics of collapse in world fisheries. Fish and Fisheries 6: 111–120.

Myers, R.A. and Worm, B. (2003). Rapid worldwide depletion of predatory fish communities. Nature 423: 280–283.

Myers, R.A., Barrowman, N.J., Hoenig, J.M. and Qu, Z. (1996a). The collapse of cod in Eastern Canada: the evidence from tagging data. ICES Journal of Marine Science 53: 629–640.

Myers, R.A., Hutchings, J.A. and Barrowman, N.J. (1996b). Hypothesis for the decline of cod in the North Atlantic. Marine Ecology Progress Series 138: 293–308.

National Research Council (2003). Understanding inputs, fates, and effects in detail. In: Oil in the Sea III: Inputs, Fates and Effects. Ch. II, pp. 63–182. The National Academies Press, Washington, DC. ISBN: 0-309-08438-5.

Naylor, R., Hindar, K., Fleming, I.A., Goldburg, R., Williams, S., Volpe, J., Whoriskey, F., Eagle, J., Kelso, D. and Mangel, M. (2005). Fugitive salmon: assessing the risks of escaped fish from net-pen aquaculture. BioScience 55(5): 427–437.

Naylor, R.L. and Burke, M. (2005). Aquaculture and ocean resources: raising tigers of the sea. Annual Review of Environment and Resources 30: 185–218.

Naylor, R.L., Goldburg, R.J., Mooney, H., Beveridge, M., Clay, J., Folke, C., Kautsky, N., Lubchenco, J., Primavera, J. and Williams, M. (1998). Nature's subsidies to shrimp and salmon farming. Science 282: 883–884.

Naylor, R.L., Goldburg, R.J., Primavera, J.H., Kautsky N., Beveridge, M.C.M., Clay, J., Folkes, C., Lubchenco, J., Mooney, H. and Troell, M. (2000). Effect of aquaculture on world fish supplies. Nature 405: 1017–1023.

Naylor, R.L., Eagle, J. and Smith, W.L. (2003). Salmon aquaculture in the Pacific Northwest. A global industry. Environment 45(8): 18–39.

Nel, D.C., Ryan, P.G., Crawford, R.J.M., Cooper, J. and Huyser, A.W. (2002). Population trends of albatrosses and petrels at sub-Antarctic Marion Island. Polar Biology 25: 81–89.

Neori, A., Chopin, T., Troell, M., Buschmann, A.H., Kraemer, G.P., Halling, C., Shpigel, M. and Yarish, C. (2004). Integrated aquaculture: rationale, evolution and state of the art emphasizing seaweed biofiltration in modern mariculture. Aquaculture 231: 361–391.

Neori, A., Troell, M., Chopin, T., Yarish, C., Critchley, A. and Buschmann, A.H. (2007). The need for a balanced ecosystem approach to blue revolution aquaculture. Environment 49(3): 36–43.

NOAA Fisheries (2006a). Turtle excluder devices (TEDS). National Oceanic and Atmospheric Administration (NOAA) Fisheries, Office of Protected Resources. Accessed 2007 at: http://www.nmfs.noaa.gov/pr/species/turtles/teds.htm

NOAA (2006b). President sets aside largest marine conservation area on earth. National Oceanic and Atmospheric Administration (NOAA) Fisheries, Office of Protected Resources. Accessed 2007 at: http://www.noaanews.noaa.gov/stories2006/s2644.htm

NOAA (2007). Marine Mammal Protection Act (MMPA) of 1972. National Oceanic and Atmospheric Administration (NOAA) Fisheries, Office of Protected Resources. Accessed 2007 at: http://www.nmfs.noaa.gov/pr/laws/mmpa/

Norse, E.A., Crowder, L.B., Gjerde, K., Hyrenbach, D., Roberts, C.M., Safina, C. and Soule, M.E. (2005). Place-based ecosystem management in the open ocean. In: Marine Conservation Biology: The Science of Maintaining the Sea's Biodiversity (eds. E. Norse and L. Crowder), pp. 302–327. Island Press, Washington, DC.

4th NSC (1995). Ministerial declaration of the Fourth International Conference on the protection of the North Sea ('The Esbjerg Declaration'), Esbjerg, Denmark. Accessed 2007 at: http://www.sweden.gov.se/content/1/c6/05/78/43/ee1a9f5d.pdf

Nugues, M.M., Smithy, G.W., van Hooidonk, R.J., Seabra, M.I. and Bak, R.P.M. (2004). Algal contact as a trigger for coral disease. Ecology Letters 7: 919–923.

Ormond, R.F.G. and Roberts, C.M. (1997). The biodiversity of coral reefs fishes. In: Marine Biodiversity: Patterns and Processes (eds. R.F.G. Ormond, J.D. Gage and M.V. Angel), Ch. 10, pp. 216–257. Cambridge University Press, Cambridge. ISBN: 0-521-55222-2.

Orr, J.C., Fabry, V.J., Aumont, O., Bopp, L., Doney, S.C., Feely, R.A., Gnanadesikan, A., Gruber, N., Ishida, A., Joos, F., Key, R.M., Lindsay, K., Maier-Reimer, E., Matear, R., Monfray, P., Mouchet, A., Najjar, R.G., Plattner, G.-K., Rodgers, K.B., Sabine, C.L., Sarmiento, J.L., Schlitzer, R., Slater, R.D., Totterdell, I.J., Weirig, M.-F., Yamanaka, Y. and Yool, A. (2005). Anthropogenic ocean acidification over the twenty-first century and its impact on calcifying organisms. Nature 437: 681–686.

OSPAR (1998a). OSPAR Strategy with Regard to Hazardous Substances. OSPAR Convention for the Protection of the Marine Environment of the North-East Atlantic, OSPAR 98/14/1 Annex 34.

OSPAR (1998b) 'The Sintra statement'. Statement from the ministerial meeting of the OSPAR Commission, Sintra, Portugal, July 1998. Accessed 2007 at: http://www.ospar.org/eng/html/md/mainresult.htm

OSPAR (1998c) OSPAR Strategy with Regard to Radioactive Substances. OSPAR Convention for the Protection of the Marine Environment of the North-East Atlantic, OSPAR 98/14/1.

OSPAR (2003). 2003 Progress report on the more detailed implementation of the OSPAR strategy with regard to radioactive substances. OSPAR Convention for the Protection of the Marine Environment of the North-East Atlantic. Bremen 25 June. Accessed 2007 at: http://www.ospar.org/documents/02-03/OSPAR03/SR-E/ANNEX30_Progress%20Report%20on%20Radioactive%20Strategy.doc

OSPAR (2004). Meeting of the OSPAR Commission, Reykjavik: 28 June–2 July 2004. Cooperation with the Bonn Agreement, presented by the Secretariat.

OSPAR Convention for the Protection of the Marine Environment of the North-East Atlantic.

OSPAR (2006). OSPAR List of Chemicals for Priority Action (update 2006) (Reference Number 2004-12). OSPAR Convention for the Protection of the Marine Environment of the North-East Atlantic.

Osvath, I., Povinec, P.P., Baxter, M.S., Huynh-Ngoc, L. (2001). Mapping of the distribution of ^{137}Cs in Irish Sea sediments. Journal of Radioanalytical and Nuclear Chemistry 248(3): 735–739.

Oxfam (2006). Fishing for a future. The advantages and drawbacks of a comprehensive fisheries agreement between the Pacific and European Union. Oxfam, New Zealand. 22 pp.

Pala, C. (2007). No-fishing zones in tropics yield fast payoffs for reefs. New York Times, 17 April 2007. Pandolfi, J.M., Bradbury, R.H., Sala, E., Hughes, T.P., Bjorndal, K.A., Cooke, R.G., McArdle, D., McClenachan, L., Newman, M.J.H., Paredes, G., Warner, R.R. and Jackson, J.B.C. (2003). Global trajectories of the long-term decline of coral reef ecosystems. Science 301: 955–960.

Pandolfi, J.M., Jackson, J.B.C., Baron, N., Bradbury, R.H., Guzman, H.M., Hughes, T.P., Kappel, C.V., Micheli, F., Ogden, J.C., Possingham, H.P. and Sala, E. (2005). Are U.S. coral reefs on the slippery slope to slime? Science 307: 1725–1726.

Paramo, J., Quiñones, R.A., Ramirez, A. and Wiff, R. (2003). Relationship between abundance of small pelagic fishes and environmental factors in the Colombian Carribbean Sea: an analysis based on hydroacoustic information. Aquatic Living Resources 16: 239–245.

Pauly, D. and Christensen, V. (1995). Primary production required to sustain global fisheries. Nature 374: 255–257.

Pauly, D. and Palomares, M.-L. (2005). Fishing down marine food web: it's far more pervasive than we thought. Bulletin of Marine Science 76(2): 197–211.

Pauly, D. and Watson, R. (2003). Counting the last fish. Scientific American 289(1): 34–39.

Pauly, D. and Watson, R. (2005). Background and interpretation of the 'marine trophic index' as a measure of biodiversity. Philosophical Transactions of the Royal Society B: Biological Sciences 360: 415–423.

Pauly, D., Christensen, V., Guénette, S., Pitcher, U., Sumaila, R., Walters, C.J., Watson, R. and Zeller, D. (2002). Towards sustainability in world fisheries. Nature 418: 689–695.

Pauly, D., Watson, R. and Alder, J. (2005). Global trends in world fisheries: impacts on marine ecosystems and food security. Philosophical Transactions of the Royal Society B: Biological Sciences 360: 5–12.

Pearson, M. and Inglis, V. (1993). A sensitive microbioassay for the detection of antibacterial agents in the aquatic environment. Journal of Fish Diseases 16: 255–260.

Peck, L.S., Webb, K.E. and Bailey, D.M. (2004). Extreme sensitivity of biological function to temperature in Antarctic marine species. Functional Ecology 18: 625–630.

Pérez, J.E., Alfonsi, C., Nirchio, M., Muñon, C. and Gómez, J.A. (2003). The introduction of exotic species in aquaculture: a solution or part of the problem? Interciencia 28(4): 234–238.

Perry, A.L., Low, P.J., Ellis, J.R. and Reynolds, J.D. (2005). Climate change and distribution shifts in marine fishes. Science 308(5730): 1912–1915.

Peterson, C.H. (2001). The "Exxon Valdez" oil spill in Alaska: acute, indirect and chronic effects on the ecosystem. Advances in Marine Biology 39: 3–103.

Peterson, C.H., Rice, S.D., Short, J.W., Esler, D., Bodkin, J.L., Ballachey, B.E. Irons, D.B. (2003). Long-term ecosystem response to the Exxon Valdez oil spill. Science 302: 2082–2086.

Phillips, R.C. and Durako, M.J. (2000). Global status of seagrasses. In: Seas at the Millennium: An Environmental Evaluation. Volume III, Global Issues and Processes, (ed. C. Sheppard), Ch. 107, pp. 1–16. Pergamon/Elsevier Science, Oxford.

Phyne, J. and Mansilla, J. (2003). Forging linkages in the commodity chain: the case of the Chilean salmon farming industry. Sociologica Ruralis 43(2): 108–127.

Pickova, J. and Mørkøre, T. (2007). Alternate oils in fish feeds. European Journal of Lipid Science and Technology 109: 256–263.

Pinedo, M.C. and Polacheck, T. (2004). Sea turtle by-catch in pelagic longline sets off southern Brazil. Biological Conservation 119: 335–339.

Pizarro, R. (2006). APP No. 37: The ethics of world food production: the case of salmon-farming in Chile. Paper presented at the Conference 'Ethics of Globalization', Cornell, 29–30 September 2006. Publicaciones Fundacion Terram, Santiago, Chile.

Prieur, D. (1997). Editorial. Biodiversity and Conservation 6(11): R1–R2.

Public Citizen (2004). Shell game. The environmental and social impacts of shrimp aquaculture. Public Citizen, Washington, DC. 20 pp.

Quinton, C.D., Kause, A., Koskela, J. and Ritola, O. (2007). Breeding salmonids for feed efficiency in current fishmeal and future plant-based diet environment. Genetics Selection Evolution 39: 431–466.

Rabalais, N., Turner, R.E. and Wiseman, W.J. (2002). Gulf of Mexico hypoxia, A.K.A."The Dead Zone". Annual Review of Ecology and Systematics 33: 235–263.

Rahmstorf, S. (2000). The thermohaline ocean circulation: a system with dangerous thresholds. Climate Change 46: 247–256.

Rahmstorf, S. (2002). Ocean circulation and climate during the past 120,000 years. Nature 419: 207–214.

Raloff, J. (2004). Dead waters: massive oxygen-starved zones are developing along the world's coasts. Science News 165(23): 360.

Ramirez-Llodra, E., Shank, T.M. and German, C.R. (2007). Biodiversity and biogeography of hydrothermal vent species. Thirty years of discovery. Oceanography 20(1): 30–41.

Raven, H. (2006). Organic salmon – setting the standard. Presentation at Aquaculture Conference, The Tolbooth, Stirling, 23 March 2006. Accessed November 2007 at: http://www.soilassociationscotland.org/

RCEP (2004). Turning the tide: addressing the impact of fisheries on the marine environment. Royal Commission on Environmental Pollution 25th Report. December 2004. 480 pp.

Read, A.J. and Rosenberg, A.A. (2002). Draft International Strategy for Reducing Incidental Mortality of Cetaceans in Fisheries. Cetacean By-catch Resource Centre. Accessed February 2008 at: http://www.cetaceanbycatch.org/intlstrategy.cfm

Read, A.J., Drinker, P. and Northridge, S. (2006). By-catch of marine mammals in U.S. and global fisheries. Conservation Biology 20(1): 163–169.

Reaser, J.K., Pomerance, R. and Thomas, P.O. (2000). Coral bleaching and global climate change: scientific findings and policy recommendations. Conservation Biology 14(5): 1500–1511.

Reeves, R.R., Smith, B.D., Crespo, E.A. and Notarbartolo di Sciara, G. (2003). Dolphins, whales and porpoises: 2002–2010 conservation action plan for the world's cetaceans. IUCN/SSC Cetacean Specialist Group. IUCN, Gland, Switzerland/Cambridge. xi + 139 pp.

Renard, Y. (2001). Case of the Soufrière Marine Management Area (SMMA), St. Lucia. The Caribbean Natural Resources Institute. CANARI Technical Report N1285. 8 pp.

Renner, R. (2001). Growing concern over perfluorinated chemicals. Environmental Science and Technology 35(7): 155A–160A.

Renner, R. (2003). Concerns over common perfluorinated surfactant. Environmental Science and Technology 37(11): 201A–202A.

Renner, R. (2004). Tracking the dirty by products of a world trying to stay clean. Science 306(5703): 1887.

Renner, R. (2006). Exxon Valdez oil no longer a threat? Environmental Science and Technology 40(20): 6188–6189.

Rijnsdorp, A.D. and van Leeuwen, P.I. (1996). Changes in growth of North Sea plaice since 1950 in relation to density, eutrophication, beam-trawl effort, and temperature. Ices Journal of Marine Science 53: 1199–1213.

Roberts, C.M. (1995). Effects of fishing on the ecosystem structure of coral reefs. Conservation Biology 9(5): 988–995.

Roberts, C.M. and Hawkins, J.P. (2000). Fully protected marine reserves: a guide. WWF Endangered Seas Campaign, 1250 24th Street, NW, Washington, DC 20037, USA and Environment Department, University of York, York, YO10 5DD, UK. ISBN: 2-88085-239-0.

Roberts, C.M., Bohnsack, J.A., Gell, F., Hawkins, J.P. and Goodridge, R. (2001). Effects of marine reserves on adjacent fisheries. Science 294: 1920–1922.

Roberts, C.M., McClean, C.J., Veron, J.E.N., Hawkins, J.P., Allen, G.R., McAllister, D.E., Mittermeier, C.G., Schueler, F.W., Spalding, M., Wells, F., Vynne, C. and Werner, T.B. (2002) Marine biodiversity hotspots and conservation priorities for tropical reefs. Science 295: 1280–1284.

Roberts, C.M., Andelman, S., Branch, G., Bustamante, R.H., Castilla, J.C., Dugan, J., Halpern B.S., Lafferty, K.D., Leslie, H., Lubchenco, J., McArdle, D., Possingham, H.P., Ruckelshaus, M. and Warner, R.R. (2003). Ecological criteria for evaluating candidate sites for marine reserves. Ecological Applications 13(1): 199–214.

Roberts, C.M., Mason, L., Hawkins, J.P., Masden, E., Rowlands, G., Storey, J. and Swift, A. (2006). Roadmap to Recovery: A Global Network of Marine Reserves. Greenpeace International, Amsterdam, The Netherlands. 58 pp.

Roberts, S. and Hirshfield, M. (2004). Deep-sea corals: out of sight, but no longer out of mind. Frontiers in Ecology and the Environment 2(3): 123–130.

Robins, J.B. (1995). Estimated catch and mortality of sea turtles from the east coast otter trawl fishery of Queensland, Australia. Biological Conservation 74: 157–167.

Roessig, J.M., Woodley, C.M., Cech, J.J. and Hansen, L.J. (2004). Effects of global climate change on marine and estuarine fishes and fisheries. Reviews in Fish Biology and Fisheries 14: 251–275.

Rogers, A.D. (1994). The biology of seamounts. Advances in Marine Biology 30: 305–350.

Rojas-Bracho, L., Reeves, R.R., and Jaramillo-Legoretta, A. (2006). Conservation of the vaquita *Phocoena sinus*. Mammal Review 36(3): 179–216.

Rönnbäck, P. (1999). The ecological basis for economic value of seafood production supported by mangrove ecosystems. Ecological Economics 29: 235–252.

Ross, A. and Isaac, S. (2004). The net effect? A review of cetacean bycatch in pelagic trawls and other fisheries in the north-east Atlantic. A WDCS report for Greenpeace. Greenpeace Environmental Trust. 73 pp.

Royal Society (2005). Ocean acidification due to increasing atmospheric carbon dioxide. The Royal Society. Policy document 12/05. ISBN: 0-85403-617-2. 58 pp.

Rudd, M.A. and Tupper, M.H. (2002). The impact of Nassau grouper size and abundance on scuba diving selection and MPA economics. Coastal Management 30(2): 133–151.

Russ, G.R., Alcala, A.C. and Maypa, A.P. (2003). Spillover from marine reserves: the case of *Naso vlamingii* at Apo Island, the Philippines. Marine Ecology Progress Series 264: 15–20.

Ryan, T.P., Dowdall, A.M., Long, S., Smith, V., Pollard, D. and Cunningham, J.D. (1999). Plutonium and americium in fish, shellfish and seaweed in the Irish environment and their contribution to dose. Journal of Environmental Radioactivity 44: 349–369.

Sadovy, Y. (2005). Trouble on the reef: the imperative for managing vulnerable and valuable fisheries. Fish and Fisheries 6: 167–185.

Sale, P.F. (1991). Introduction. In: The Ecology of Fishes on Coral Reefs (eds. P.F. Sale), Ch. 1, pp. 3–15. Academic, San Diego, CA. ISBN: 0-12-615180-6.

Sánchez, F., Velasco, F., Cartes, J.E., Olaso, I., Preciado, I., Fanelli, E., Serrano, A. and Gutierrez-Zabala, J.L. (2006). Monitoring the Prestige oil spill impacts on some key species of the Northern Iberian shelf. Marine Pollution Bulletin 53(5–7): 332–349.

Sánchez, P., Demestre, M. and Martín, P. (2004). Characterisation of the discards generated by bottom trawling in the northwestern Mediterranean. Fisheries Research 67(1): 71–80.

Santiago Times (2007). Unions scrutinize labor problems in Chile's salmon industry. 5 December 2007.

Santillo, D., Allsopp, M., Walters, A., Johnston, P. and Perivier, H. (2006). The presence of PFOS and other perfluorinated chemicals in eels (*Anguilla anguilla*) from 11 European countries. Greenpeace International. Greenpeace Research Laboratories Technical Note 07/2006. 31 pp.

Sarkar, S.K. and Bhattacharya, A.K. (2003). Conservation of biodiversity of coastal resources of Sundarbans, Northeast India: and integrated approach through environmental education. Marine Pollution Bulletin 47: 260–264.

Scheffer, M., Carpenter, S. and de Young, B. (2005). Cascading effects of overfishing marine systems. Trends in Ecology and Evolution 20(11): 579–581.

Schrank, W.E. (2005). The Newfoundland fishery: ten years after the moratorium. Marine Policy 29: 407–420.

Schratzberger, M. and Jennings, S. (2002). Impacts of chronic trawling disturbance on meiofaunal communities. Marine Biology 141: 991–1000.

Schultz, M.M., Barofsky, D.F. and Field, J.A. (2004). Quantitative determination of fluorotelomer sulfonates in groundwater by LC MS/MS. Environmental Science and Technology 38(6): 1828–1835.

Scottish Executive Central Research Unit (2002). Review and synthesis of the environmental impacts of aquaculture. The Scottish Association for Marine Science and Napier University. Scottish Executive Central Research Unit. The Stationery Office, Edinburgh, UK. 71 pp.

SCRS (2006). Report of the 2006 Atlantic bluefin tuna assessment. Session of the Scientific Committee on Research and Statistics. SCRS /2006/013. International Commission for the Conservation of Atlantic Tunas.

Sebens, K.P. (1994). Biodiversity of coral reefs: what are we losing and why? American Zoologist 34: 115–133.

Secretariat of the Convention on Biological Diversity (2003). Interlinkages between biological diversity and climate change. Advice on the integration of biodiversity considerations into the implementation of the United Nations Framework Convention on Climate Change and its Kyoto Protocol. Montreal, SCBD (CBD Technical Series No. 10). 154 pp.

Sellström, U., Bignert, A., Kierkegaard, A., Häggberg, L., de Wit, C.A., Olsson, M. and Jansson, B. (2003). Temporal trend studies on tetra- and pentabrominated diphenyl ethers and hexabromo-cyclododecane in guillemot egg from the Baltic Sea. Environmental Science and Technology 37(24): 5496–5501.

Serfling, S.A. (2006). Use of suspended microbial treatment methods in recirculating tank systems, for culture of freshwater and marine fish, and marine shrimp. Presented at: WAS America Meeting, Las Vegas. Microbial Controlled Systems. Special symposium, February 15 2006.

She, J., Petreas, M., Winkler, J., Visita, P., McKinney, M. and Kopec, D. (2002). PBDEs in the San Francisco Bay area: measurements in harbour seal blubber and human breast adipose tissue. Chemosphere 46: 697–707.

Shepherd, C.J., Pike, I.H. and Barlow, S.M. (2005). Sustainable feed resources of marine origin. Presented at Aquaculture Europe 2005. European Aquaculture Society Special Publication No. 35. June 2005. pp. 59–66.

Shoeib, M., Harner, T., Wilford, B., Jones, K. and Zhu, J. (2004). A survey of perfluoroalkyl sulfonamides in indoor and outdoor air using passive air samplers. Organohalogen Compounds 66: 3999–4003.

Short, F.T. and Neckles, H.A. (1999). The effects of global climate change on seagrasses. Aquatic Botany 63: 169–196.

Short, J.W., Rice, S.D., Heinz, R.A., Carls, M.G. and Moles, A. (2003). Long-term effects of crude oil on developing fish: lessons from the Exxon Valdez oil spill. Energy Sources 25: 509–517.

Short, J.W., Lindeberg, M.R., Harris, P.M., Maselko, J.M., Pella, J.J. and Rice, S.D. (2004). Estimate of oil persisting on the beaches of Prince William Sound 12 years after the Exxon Valdez oil spill. Environmental Science and Technology 38: 19–25.

Short, J.W., Maqselko, J.M., Lindeberg, M.R., Harris, P.M. and Rice, S.D. (2006). Vertical distribution and probability of encountering intertidal Exxon Valdez oil on shorelines of three embayments within Prince William Sound Alaska. Environmental Science and Technology 40(12): 3723–3729.

Silvani, L., Gazo, M. and Aguiler, A. (1999). Spanish driftnet fishing and incidental catches in the western Mediterranean. Biological Conservation 90: 79–85.

Singkran, N. and Sudara, S. (2005). Effects of changing environments of mangrove creeks on fish communities at Trat Bay, Thailand. Environmental Management 35(1): 45–55.

Slooten, E., Dawson, S., Rayment, W. and Childerhouse, S. (2006). A new abundance estimate for Maui's dolphin: what does it mean for managing this critically endangered species. Biological Conservation 128: 576–581.

Small, C.J. (2005). Regional fisheries management organisations: their duties and performance in reducing by-catch of albatrosses and other species. Birdlife Global Seabird Programme, Birdlife International. Cambridge. 103 pp.

Smetacek, V. and Nicol, S. (2005). Polar ocean ecosystems in a changing world. Nature 437: 362–368.

Smith, R.C., Ainley, D., Baker, K., Domack, E., Emslie, S., Fraser, B., Kennett, J., Leventer, A., Mosley-Thomson, E., Stammerjohn, S. and Vernet, M. (1999). Marine ecosystem sensitivity to climate change. Bioscience 49(5): 393–404.

Smithwick, M., Mabury, S.A., Solomon, K.R., Sonne, C., Martin, J.W., Born, E.W., Dietz, R., Derocher, A.E., Letcher, R.J., Evans, T.J., Gabrielsen, G.W., Nagy, J., Stirling, I., Taylor, M.K. and Muir, D.C.G. (2005). Circumpolar study of perfluoroalkyl contaminants in polar bears (*Ursus maritimus*). Environmental Science and Technology 39(15): 5517–5523.

Smithwick, M., Norstrom, R.J., Mabury, S.A., Solomon, K., Evans, T.J., Stirling, I. and Muir, D.C.G. (2006). Temporal trends of perfluoroalkyl contaminants in polar bears (*Ursus maritimus*) from two locations in the North American Arctic, 1972–2002. Environmental Science and Technology 40(4): 1139–1143.

So, M.K., Taniyasu, S., Yamashita, N., Giesy, J.P., Zheng, J., Fang, Z., Im, S.H. and Lam, P.K.S. (2004). Perfluorinated compounds in coastal waters of Hong Kong, South China, and Korea. Environmental Science and Technology 38(15): 4056–4063.

Sobel, J. and Dahlgren, C. (2004). Marine Reserves: A Guide to Science, Design and Use. Island Press, Washington, DC. 220 pp. ISBN: 1-55963-841-9.

Spalding M.D., Ravilious C. and Green E.P. (2001). World Atlas of Coral Reefs. Prepared at the UNEP World Conservation Monitoring Centre. University of California Press, Berkeley, CA. 424 pp.

Stapleton, H.M. (2006). Brominated flame retardants: assessing DecaBDE debromination in the environment. EPHA Environment Network (EEN), Brussels. 9 pp.

Stock, N.L., Lau, F.K., Ellis, D.A., Martin, J.W., Muir, D.C.G. and Mabury, S.A. (2004). Polyfluorinated telomer alcohols and sulfonamides in the North American troposphere. Environmental Science and Technology 38(4): 991–996.

Stockholm Convention on Persistent Organic Pollutants (POPs) (2008). Accessed February 2008 at: http://www.pops.int/

Stocks, K. (2004a). Seamounts online: an online resource for data on the biodiversity of seamounts. In: Seamounts: Biodiversity and Fisheries (eds. T. Morato and D. Pauly). Fisheries Centre Research Reports 12(5): 13–16. Fisheries Centre, University of British Columbia, Canada.

Stocks, K. (2004b). Seamount invertebrates: composition and vulnerability to fishing. In: Seamounts: Biodiversity and Fisheries (eds. T. Morato and D. Pauly). Fisheries Centre Research Reports 12(5): 17–24. Fisheries Centre, University of British Columbia, Canada.

Stoker, T.E., Cooper, R.L., Lambright, C.S., Wilson, V.S., Furr, J. and Gray, L.E. (2005). In vivo and in vitro anti-androgenic effects of DE-71, a commercial polybrominated diphenyl ether (PBDE) mixture. Toxicology and Applied Pharmacology 207: 78–88.

Sugiyama, S., Staples, D. and Funge-Smith, S.J. (2004). Status and Potential of Fisheries and Aquaculture in Asia and the Pacific. RAP Publication 2004/25. FAO Regional Office for Asia and the Pacific, Bangkok. 53 pp.

Swedish Chemicals Inspectorate KemI (2005). Proposal for listing Perfluorooctane sulfonate (PFOS) in Annex A of the Stockholm Convention on Persistent Organic Pollutants.

Szmant, A.M. (2002). Nutrient enrichment on coral reefs: is it a major cause of coral reef decline? Esturaries 25(4b): 743–766.

Tacon, A.G.J. (2005). State of information on salmon aquaculture feed and the environment. Report prepared for the WWF US initiated salmon aquaculture dialogue. 80 pp.

Tacon, A.G.J., Hasan, M.R. and Subasinghe, R.P. (2006). Use of Fishery Resources as Feed Inputs for Aquaculture Development: Trends and Policy Implications. FAO Fisheries Circular. No. 1018, Food and Agricultural Organization of the United Nations, Rome. 99 pp.

Tada, Y., Fujitani, T., Ogata, A. and Kamimura, H. (2007). Flame retardant tetrabromobisphenol A induced hepatic changes in ICR male mice. Environmental Toxicology and Pharmacology 23(2): 174–178.

Taniyasu, S., Kannan, K., Horii, Y., Hanari, N. and Yamashita, N. (2003). A survey of perfluoroocatane sulfonate and related perfluorinated organic compounds in water, fish, birds, and humans from Japan. Environmental Science and Technology 37(12): 2634–2639.

Taniyasu, S., Yamashita, N., Kannan, K., Horii, Y., Sinclai, E., Petrick, G. and Gamo, T. (2004). Perfluorinated carboxylates and sulfonates in open ocean waters of the Pacific and Atlantic Oceans. Organohalogen Compounds 66: 4035–4039.

Thacker, R.W., Ginsburg, D.W. and Paul, V.J. (2001). Effects of herbivore exclusion and nutrient enrichment on coral reef macroalgae and cyanobacteria. Coral Reefs 19: 318–329.

Tidens Krav (2007). Krever oppdrettsnæringa. Accessed January 2008 at: http://www.tk.no/nyheter/article2502556.ece

Tracey, D.M., Bull, B., Clark, M.R. and Mackay, K.A. (2004). Fish species composition on seamounts and adjacent slope in New Zealand waters. New Zealand Journal of Marine and Freshwater Research 38: 163–182.

Tuck, G.N., Polacheck, T. and Bulman, C. (2003). Spatio-temporal trends of longline fishing effort in the Southern Ocean and implications for seabird by-catch. Biological Conservation 114: 1–27.

Ueno, D., Kajiwara, N., Tanaka, H., Subramanian, A., Fillmann, G., Lam, P.K.S., Zheng, G.J., Muchitar, M., Razak, H., Prudente, M., Chung, K-H. and Tanabe, S. (2004). Global pollution monitoring of polybrominated diphenyl ethers using skipjack tuna as a bioindicator. Environmental Science and Technology 38: 2312–2316.

UNEP (1995). Global Biodiversity Assessment. United Nations Environment Programme (UNEP). Cambridge University Press, Cambridge. ISBN: 0-521-56403-4. 1140 pp.

UNEP (2006a). Ecosystems and biodiversity in deep waters and high seas. UNEP Regional Seas Report and Studies No. 178. UNEP/IUCN, Switzerland 2006. ISBN: 92-807-2734-6.

UNEP (2006b). After the Tsunami: Rapid Environmental Assessment. United Nations Environment Programme. 142 pp.

UNEP (2006c). Further rise in number of marine 'dead zones'. Press release 19 October 2006. Accessed 2007 at: http://www.unep.org/Documents.Multilingual/Default.asp?DocumentID= 486&ArticleID=5393&l=en

UNEP-WCMC (2006). In the front line: shoreline protection and other ecosystem services from mangroves and coral reefs. UNEP-WCMC (United Nations Environment Programme – World Conservation Monitoring Centre), Cambridge. 33 pp.

UNEP and WHRC (2007). Reactive nitrogen in the environment: too much or too little of a good thing. United Nations Environment Programme, Paris.

UNFCCC (2005). Climate Change, Small Island Developing States. Issued by the Climate Change Secretariat (United Nations Framework Convention on Climate Change, UNFCCC), Bonn, Germany. ISBN: 92-9219-012-1. 32 pp.

US EPA (2006a). Perfluorooctanoic acid. Accessed 2007 at: http://www.epa.gov/oppt/pfoa/

US EPA (2006b). EPA seeking PFOA reductions. Press release 25 January 2006. Accessed 2007 at: http://yosemite.epa.gov/opa/admpress.nsf/68b5f2d54f3eefd28525701500517fbf/fd1cb3a0 75697aa485257101006afbb9!opendocument

Valentine, J.F. and Heck, K.L. Jr. (2005). Perspective review of the impacts of overfishing on coral reef food web linkages. Coral Reefs 24: 209–213.

Valiela, I., Bowen, J.L. and York, J.K. (2001). Mangrove forests: one of the world's threatened major tropical environments. Bioscience 51(10): 807–815.

Van de Vijver, K.I., Hoff, P.T., Das, K., Van Dongen, W., Esmans, E.L., Jauniaux, T., Bouquegneau, J.M., Blust, R. and De Coen, W.M. (2003). Perfluorinated chemicals infiltrate ocean waters: link between exposure levels and stable isotope ratios in marine mammals. Environmental Science and Technology 37(24): 5545–5550.

Van de Vijver, K.I., Hoff, P.T., Das, K., Van Dongen, W., Esmans, E.L., Siebert, U., Bouquegneau, J.M., Blust, R. and De Coen, W.M. (2004). Baseline study of perfluorochemicals in harbour porpoises (Phocoena phocoena) from Northern Europe. Marine Pollution Bulletin 48(9–10): 992–1008.

Van Dover, C. (2005). Hot topics: biogeography of deep-sea hydrothermal vent faunas. Accessed 2007 at: http://www.divediscover.whoi.edu/hottopics/biogeo.html

Velando, A., Munilla, I. and Leyenda, P.M. (2005). Short-term indirect effects of the 'Prestige' oil spill on European shags: changes in availability of prey. Marine Ecology Progress Series 302: 263–274.

Verdegem, M.C.J., Schrama, J., Hari, B. and Kurup, M. (2006). Microbial controlled production of P. monodon in extensive ponds. Presented at: WAS America Meeting, Las Vegas. Microbial Controlled Systems. Special symposium, 15 February 2006.

Viberg, H., Johansson, N., Fredriksson, A., Eriksson, J., Marsh, G. and Erikksson, P. (2006). Neonatal exposure to higher brominated diphenyl ethers, hepta, octa-, or nonabromodiphe-nyl ether, impairs spontaneous behaviour and learning and memory functions of adult mice. Toxicological Sciences 92(1): 211–218.

Vidal, J. (2005). Pacific Atlantis: first climate change refugees. The Guardian, Firday 25 November 2005.

Volpe, J.P. (2005). Dollars without sense: the bait for big-money tuna ranching around the world. BioScience 55(4) (April): 301–302.

Vorkamp, K., Dam, M., Riget, F., Fauser, P., Bossi, R. and Hansen, A.B. (2004a). Screening of "new" contaminants in the marine environment of Greenland and the Faroe Islands. National Environmental Research Institute, Denmark. NERI Technical Report No 525. 97 pp.

Vorkamp, K., Christensen, J.H., Glasius, M. and Riget, F.F. (2004b). Persistent halogenated com-pounds in black guillemots (Cepphus grille) from Greenland – levels, compound patterns and spatial trends. Marine Pollution Bulletin 48: 111–121.

Wabnitz, C., Taylor, M., Green, E. and Razak, T. (2003). From Ocean to Aquarium. UNEP World Conservation Monitoring Centre, Cambridge. 64 pp.

Walters, A. and Santillo, D. (2006) Uses of perfluorinated substances. Greenpeace Research Laboratories. Technical Note 06/2006. 20 pp.

Ward, T.M., McLeay, L.J., Dimmlich, W.F., Rogers, P.J., McClatchie, S., Matthews, R., Kämpf, J. and Van Ruth, P.D. (2006). Pelagic ecology of a northern boundary current system: effects of upwelling on the production and distribution of sardine (*Sardinops sagax*), anchovy (*Engraulis australis*) and southern bluefin tuna (*Thunnus maccoyii*) in the Great Austrailian Bight. Fisheries Oceanography 15(3): 191–207.

Wasielesky, W., Atwood, H., Stokes, A. and Browdy, C.L. (2006). Effect of natural production in a zero exchange suspended microbial floc super-intensive culture system for white shrimp *Litopenaeus vannamei*. Aquaculture 258(1–4): 396–403.

Watling, L. and Norse, E.A. (1998). Disturbance of the seabed by mobile fishing gear: a comparison to forest clearcutting. Conservation Biology 12(6): 1180–1197.

Watson, J.W., Epperly, S.P., Shah, A.K. and Foster, D.G. (2005). Fishing methods to reduce sea turtle mortality associated with pelagic longlines. Canadian Journal of Fisheries and Aquatic Sciences 62(5): 965–981.

Watson, W.S., Sumner, D.J., Baker, J.R., Kennedy, S., Reid, R. and Robinson, I. (1999). Radionuclides in seals and porpoises in coastal waters around the UK. The Science of the Total Environment 234: 1–13.

Weimerskirch, H., Catard, A., Prince, P.A., Cherel, Y. and Croxall, J.P. (1999). Foraging white-chinned petrels *Procellaria aequinoctialis* at risk: from the tropics to Antarctica. Biological Conservation 87: 273–275.

Weller, G. (2004). Summary and synthesis of the ACIA. In: ACIA (Arctic Climate Impact Assessment) Scientific Report. Ch. 18, pp. 989–1020. Cambridge University Press, Cambridge.

White, A.T., Vogt, H.P. and Arin, T. (2000). Philippine coral reefs under threat: the economic losses caused by reef destruction. Marine Pollution Bulletin 40(7): 598–605.

Williamson, D.H., Russ, G.R. and Ayling, A.M. (2004). No-take marine reserves increase abundance and biomass of reef fish on inshore fringing reefs of the Great Barrier Reef. Environmental Conservation 31(2): 149–159.

Wolff, W.J. (2000). The south-eastern North Sea: losses of vertebrate fauna during the past 2000 years. Biological Conservation 95: 209–217.

Wong, P.P. (2003). Where have all the beaches gone? Coastal erosion in the tropics. Singapore Journal of Tropical Geography 24(1): 111–132.

World Parks Congress Recommendation 22 (2003). Building a Global System of Marine and coastal protected area networks. World Parks Congress, The World Conservation Union (IUCN).

Worm, B. and Myers, R.A. (2004). Managing fisheries in a changing climate. Nature 429: 15.

Worm, B., Sandow, M., Oschlies, A., Lotze H.K. and Myers, R.A. (2005). Global patterns of predator diversity in the open oceans. Science 309: 1365–1369.

Wu, R.S.S. (1995). The environmental impact of marine fish culture – towards a sustainable future. Marine Pollution Bulletin 31(4–12): 159–166.

WWF (2008). Increasing protection: the deep sea. Accessed February 2008 at: http://www.panda. org/about_wwf/what_we_do/marine/our_solutions/protected_areas/increasing_protection/ deep_seas/index.cfm#deepsea_mpa

Yamashita, N., Kannan, K., Taniyasu, S., Horii, Y., Petrick, G. and Gamo, T. (2005). A global survey of perfluorinated acids in oceans. Marine Pollution Bulletin 51: 658–668.

Yang, Q., Xie, Y., Eriksson, A.M., Nelson, B.D. and DePierre, J.W. (2001). Further evidence for the involvement of inhibition of cell proliferation and development in thymic and splenic atrophy induced by the peroxisome proliferator perfluoroctanoic acid in mice. Biochemical Pharmacology 62: 1133–1140.

Yang, Q., Abedi-Valugerdi, M., Xie, Y., Zhao, X.-Y., Möller, G., Nelson, B.D. and DePierre, J.W. (2002). Potent suppression of the adaptive immune response in mice upon dietary exposure to the potent peroxisome proliferator perfluorooctanoic acid. International Immunopharmacology 2: 389–397.

Zegers, B.N., Mets, A., Van Bommel, R., Minkenberg, C., Hamers, T., Kamstra, J.H., Pierce, G.J. and Boon, J.P. (2005). Levels of hexabromocyclododecane in harbour porpoises and common dolphins from Western European Seas, with evidence for stereoisomer-specific biotransfor-mation by cytochrome P450. Environmental Science and Technology 39: 2095–2100.

Zeller, D. and Pauly, D. (2005). Good news, bad news: global fisheries discards are declining, but so are total catches. Fish and Fisheries 6: 156–159.

Zuberogoitia, I., Martínez, J.A., Iraeta, A., Azkona, A., Zabala, J., Jiménez, B., Merino, R. and Gómez, G. (2006). Short-term effects of the Prestige oil spill on the peregrine falcon (*Falco peregrinus*). Marine Pollution Bulletin 52(10): 1176–1181.

Index